SpringerBriefs in Environmental Science

SpringerBriefs in Environmental Science present concise summaries of cutting-edge research and practical applications across a wide spectrum of environmental fields, with fast turnaround time to publication. Featuring compact volumes of 50 to 125 pages, the series covers a range of content from professional to academic. Monographs of new material are considered for the SpringerBriefs in Environmental Science series.

Typical topics might include: a timely report of state-of-the-art analytical techniques, a bridge between new research results, as published in journal articles and a contextual literature review, a snapshot of a hot or emerging topic, an in-depth case study or technical example, a presentation of core concepts that students must understand in order to make independent contributions, best practices or protocols to be followed, a series of short case studies/debates highlighting a specific angle.

SpringerBriefs in Environmental Science allow authors to present their ideas and readers to absorb them with minimal time investment. Both solicited and unsolicited manuscripts are considered for publication.

More information about this series at http://www.springer.com/series/8868

Tejal Kanitkar

An Integrated Framework for Energy-Economy-Emissions Modeling

A Case Study of India

 Springer

Tejal Kanitkar
School of Natural Sciences and Engineering
National Institute of Advanced Studies
Bengaluru, India

ISSN 2191-5547   ISSN 2191-5555   (electronic)
SpringerBriefs in Environmental Science
ISBN 978-3-030-18262-5   ISBN 978-3-030-18263-2   (eBook)
https://doi.org/10.1007/978-3-030-18263-2

This Springer imprint is published by the registered company Springer Nature Switzerland AG.
The registered company address is: Gewerbestrasse 11, 6330 Cham, Switzerland

Abstract

The main contribution of the work presented in this brief is the integration of three modeling methods into one framework in order to address a range of questions that are of importance in the energy-economy-environment domain. Less developed countries especially the larger emerging economies such as India, China, South Africa, Brazil, and Indonesia are placed in a challenging situation. While they are growing rapidly with production in secondary and tertiary sectors fast overtaking primary agricultural activities, these countries still continue to face many challenges of ensuring the material well-being of large sections of their populations and providing access to basic services such as health, education, housing and energy. These challenges have been made more daunting by the realization of the impacts of climate change and the need to contribute to its mitigation. In this situation, these countries have to balance multiple requirements and constraints. Unlike developed countries, the economic systems in the less developed countries are still changing structurally at a very rapid pace. This makes long-range forecasts of economic variables extremely uncertain. On the other hand, to ensure that irreversible climate change does not happen, a certain amount of long-term energy planning is required. It is the task of mathematical models to provide a way of reconciling these different, often conflicting requirements.

A review of the range of such existing models, including both large macroeconomic models and technology-explicit energy models, shows some gaps in answering key questions that may be relevant to developing countries. Technology-explicit energy models such as MarkAl and TIMES are built specifically to address questions relevant to the technological aspects of the energy sector and are not suitable for addressing questions about the economy in general and about incomes and income distributions within the economy in particular. On the other hand computable general equilibrium (CGE) models consider the economy as a whole and model it using aggregated production functions and therefore can neither capture the details of the technological energy sector nor capture the full scope of the details of the specific intermediate and end use sectors of the economy. In the work presented in this report, therefore, the attempt has been to integrate three different modeling methodologies into one framework, called the integrated modeling framework (IMF) that can address some of these gaps and attempt to answer a range of questions that may be particularly important for developing countries.

The work presented here includes three main components: (a) Decomposition analysis of the past trend in emissions intensity of GDP, into its parametric components—primarily energy intensity of GDP and structural composition of the economy. Based on past trends for these factors, more robust scenarios can be constructed for the future. This component provides the basis on which scenarios are constructed in the next two components of the IMF. (b) The second model is built to evaluate the optimum energy supply pathways under various resource and emissions constraints, using simple linear programming techniques. The objective of this component is to develop a methodology for determining energy options, specifically for the power sector, under a variety of scenarios that capture the technical and economic characteristics of the power sector. (c) The third model is built to evaluate the impact of energy policy on income and equity using input-output analysis. The main objective in this part of the work is to build a methodology through which trade-offs associated with policy decisions in the power sector can be understood and quantified.

Keywords: Energy modeling; Energy-economy-emissions models; Integrated modeling framework; Input-output analysis; Structural path analysis; Decomposition analysis; Optimization; General algebraic modeling system

Nomenclature

a	Upper limit on new installed capacity that can be added annually
A	Matrix of inter-industry technical coefficients
AEM_T	Annual emissions from the power sector in year T
Anncost_t	Annual cost for period t
AP_F	Cumulative potential for each fuel
b	Lower limit on new installed capacity that can be added or decommissioned annually
BES_T	Total base load contribution to electricity generation in year (T)
BY	Base year
C	Matrix of endogenous final expenditure coefficients
CB	Carbon budget (cumulative emissions) between 2010 and 2050
CC	Capital cost of energy supply over each energy source (F) and each year (T)
CF_{ij}	Gross capital formation in sector i
d	Discount rate for each period
D_T	Demand for electricity in year T
DC_{ij}	Total depreciation
DTH_{ij}	Tax paid by household classes to the government
E_i	Energy use in sector i
EC	Row vector of energy coefficients
EF_F	Emission factor for each power generation technology (in tC/kWh)
$\text{EG}_{F,T}$	Electricity supplied by existing plants in year T
EM_i	Emissions from sector i
EP	Row vector of physical energy consumption by each sector
$\text{ES}_{F,T}$	Electricity supplied by each technology in year T
EX_{ij}	Amount of output from sector i that is exported
f_{ij}	Total final demand of commodity i in sector i
$\bar{f}$	$(n \times 1)$ column vector of exogenous final demand
$\text{FC}_{F,T}$	Fixed cost

FP_h	Cumulative population share of household group h
GA_{ij}	Net capital transfer to the government from abroad
GC_{ij}	Final consumption by government of output from sector i
GDP	Gross domestic product in real terms, not corrected for inflation unless otherwise specified
GE_{ij}	Direct monetary transfer from the government to households
GINI	Coefficient of inequality
h	Vector of exogenous household income
H	Matrix of endogenous coefficients for distributing institution and household income
H_{ij}	Total endowment (income) of households
HA_{ij}	Monetary transfer received by household classes from abroad
HC_{ij}	Final household consumption of all household classes of output from sector i
I	Actual injection into the sector
IA_{ij}	Factor income earned from abroad by capital and labor
$IC_{F,T}$	Installed capacity for each technology in year T
IG_{ij}	Total income of the government from public enterprises
IM_{jj}	Value of input in sector j that is imported
IPr_{ij}	Interest on debt paid by private corporations to the government
ITH_{ij}	Indirect tax paid by household classes in purchase of commodities
L_i	Labor employed in each sector
LC	Labor or employment coefficients
M_A	$(n \times n)$ matrix of accounting multipliers
M_1	$(n \times n)$ square matrix of direct effect multipliers
M_2	$(n \times n)$ square matrix of indirect effect multipliers
M_3	$(n \times n)$ square matrix of cross or closed loop multipliers
M_1'	$M_1 - I$
M_2'	$(M_2 - I) \times M_1$
M_3'	$(M_3 - I) \times M_2 \times M_1$
n_h	Population share of household group h
NPER	Number of periods in the planning horizon
NYRS	Number of years in each period t
OBJ	Objective function
PC_t	Per capita consumption at time t
$PLF_{F\,T}$	Plant load factor or capacity factor for each technology in year T
POP_t	Population at time t
PPr_{ij}	Total operating profit of private corporations
PPu_{ij}	Total operating surplus of public enterprises
PuS_{ij}	Savings account of public enterprises
Q	$(n \times n)$ square matrix of inter-industry technical coefficients and endogenous coefficients for distributing institution and household income

R	$(n \times n)$ square matrix of endogenous final expenditure coefficients, endogenous value-added input shares, and endogenous coefficients distributing income to value-added categories
RH_h	Rural households (h ranging from 1 to 5—poorest to richest)
RM_{ij}	Total purchase of raw material from sectors i for producing one unit of output in sector j
SA_{ij}	Total foreign savings
S	$(n \times n)$ matrix of average expenditure propensities
SH_{ij}	Total savings of household classes
SG_{ij}	Total savings account of the government
SI	Scenario for GDP growth dominated by industrial sector growth
$SPrS_{ij}$	Savings account of private corporations
SS	Scenario for GDP growth dominated by service sector growth
TC_{ij}	Tax paid on investment goods
TDSC	Total discounted system cost
TE_{ij}	Tax paid in exports
TG_{ij}	Total tax accruing to government on account of purchase of commodities by government
TIG_{ij}	Tax paid on the purchase of intermediate goods by sector j
Tin_{ij}	Total indirect tax accruing to the government
TPr_{ij}	Corporate taxes paid by private corporations to the government
TY	Target year
UH_h	Urban households (h ranging from 1 to 5—poorest to richest)
v	Vector of total value added
V	Matrix of endogenous value-added input shares
VA_{ij}	Total value added by labor and capital for production of one unit in sector j
$VC_{F,T}$	Variable cost
w	Vector of exogenous value-added income
X1	Scenario where the fraction of additional government expenditure on power comes from all sectors in equal proportion to their contribution to GDP
X2	Scenario where the fraction of additional government expenditure on power comes from a few key sectors
x	Vector of total outputs
x_i	Total production in sector i
$\bar{x}$	$(n \times 1)$ column vector of total output of all n endogenous accounts
Y	Matrix of endogenous coefficients distributing income to value-added categories
Y_{ij}	Gross output of the sector i distributed across each sector j
y_h	Income share of household group h
y	Vector of total household income
Z	Original square matrix of inter-industry transactions
Z1	Scenario with no restriction on emissions

Z2	Scenario with restriction on emissions
3E	Energy-economy-environment (used in the context of models)
ΔCF_i	Change in capital formation in between base and target years
$\Delta CF_{i\,=\,power}$	Capital formation in the power sector
ΔEI_i	Change in energy intensity if GDP for each sector
ΔG_i	Change in sectoral contribution to GDP
ΔGE_i	Change in government expenditure in between base and target years
ΔI	Total new capital formation in the power sector between base and target years

Greek Letters

α	Total factor productivity improvement
β	Efficiency improvement index
ε	Elasticity of electricity demand to price
Φ_g	Fraction of total investment that comes from the government budget
φ_{ig}	Fraction of government expenditure on the power sector that is deducted/sourced from sector i
η_h	Cumulative income share of household group h

Subscripts

F	Fuel technology
t	Time period
T	Time period
i	Input sectors in the economy
j	Output sectors in the economy

Abbreviations

AEEI	Autonomous Energy Efficiency Index
AMDI	Additive Mean Divisia Index
CGE	Computable general equilibrium
EFOM	Energy flow optimization model
FYP	Five-year plan
GAMS	General algebraic modeling system
GDP	Gross domestic product
GHG	Greenhouse gas
GNI	Gross national income
HDI	Human Development Index
IAM	Integrated assessment models

ICOR	Incremental capital output ratio
IMF	Integrated modeling framework
I-O	Input-output
IPAT	Impacts (I) of population (P), affluence (A), and technology (T)
LES	Linear expenditure system
LMDI	Log Mean Divisia Index
MarkAl	Market allocation model
NAMA	Nationally appropriate mitigation actions
RE	Renewable energy
SAM	Social accounting matrix
STIRPAT	Stochastic impacts by regression of population (P), affluence (A), and technology (T)
TFPG	Total factor productivity growth
TIMES	The Integrated MarkAl/EFOM System
T&D	Transmission and distribution
UNFCCC	United Nations Framework Convention on Climate Change

Contents

List of Figures

List of Tables

Chapter 1
Introduction

Abstract In this chapter a brief history of the development of energy studies is discussed. With a brief overview of the eras in history that characterized particular aspects of interest in energy systems and harnessing energy in its varied forms, this chapter establishes the route that energy studies have taken to arrive at the point in time today where climate change, resource availability, and capital scarcity primarily drive all investigation in energy systems. Approaches adopted to understand the complex linkages between energy systems, economic systems, and the environment are briefly discussed in this chapter before being addressed in more detail in the chapter on the review of literature.

Keywords Energy-economy-environmental linkages · Energy modeling · Climate change · Energy substitution

The production and use of modern sources of energy have gradually increased all over the world, although a substantial section of the global population mostly in the less developed countries still relies on traditional sources of energy. Many less developed countries are in the process of transitioning from traditional to modern sources of energy. However, there is an increasing recognition of constraints on the use of fossil fuels, which currently form a major share of these modern energy sources. The rapid deployment of the conventional modern energy sources such as coal, hydro, and nuclear energy, especially in the less developed countries, has been in the past constrained by the availability of capital or technological capabilities. Since the early 1990s, concerns about global warming and climate change caused by excessive emissions of greenhouse gases into the atmosphere attributable largely to the use of fossil fuels pose an additional constraint on the deployment of conventional technologies, especially ones that use coal. For less developed countries these constraints pose a challenge as they imply that they will have to meet their developmental goals without the benefit of reliance on fossil fuels to the extent that was available to the developed countries. The study of the interlinkages between energy

T. Kanitkar, *An Integrated Framework for Energy-Economy-Emissions Modeling*, SpringerBriefs in Environmental Science, https://doi.org/10.1007/978-3-030-18263-2_1

economy and the environment therefore has assumed greater importance in the era of climate change.

1.1 Background and Context

Historically, the world has seen four major energy transitions (Smil 2004). Smil classifies the harnessing of fire and the domestication of draft animals for work as the first energy transition when humans started using forms of energy that were not just somatic. Subsequently, the use of inanimate prime movers can be classified as the second energy transition when windmills and water wheels were being put to use in some parts of the world. The next major energy transition responsible for creating the modern world as we know it is the use of engines fueled by fossil fuels. The use of hydropower and more recently nuclear energy to generate electricity is also part of this transition. This transition, although more or less complete in developed countries, is still underway in a majority of the less developed countries. The final transition from the use of direct energy to indirect energy, in theory facilitating the move to cleaner forms of energy, is still underway even in developed countries. Less developed countries such as India are still in the process of completing the third energy transition and therefore the challenges of undertaking the fourth energy transition without having completed the third are particularly daunting.

Before discussing the particular challenges before developing countries, it is necessary to emphasize that although economists have worried about constraints on resources (the most notable examples being Ricardo and Malthus), the concerns have generally been about increasing pressures on food production and other resources such as water and land. Concerns about the finite nature of mineral resources, environmental pollution, and global warming are relatively more recent. Most of the issues associated with energy in the time period before 1970, especially in capital-scarce developing countries, were related to increasing access to energy and building energy infrastructure under constraints on the availability of capital. The oil shock of 1973 and the oil crisis of 1979 gave rise to increased academic interest on forecasting energy consumption and resource use. Uncertainty about extraction of mineral resources leading to large fluctuations in energy prices as well as a growing acknowledgement of the finiteness of fossil fuel resources such as oil were the main focuses of many studies published during this time. The study commissioned by the Club of Rome on the Limits to Growth (Meadows et al. 1972) was one of the most popular studies on long-range forecasting of resource consumption.

On the other hand, the United Nations conference on human environment held in Stockholm in 1972 (UN 1972) also brought to the fore concerns about local environmental pollution. The World Commission on Environment and Development, popularly known as the Brundtland Commission, in 1987 published a report that urged the world to pursue a path of sustainable development and served to re-establish the focus of many energy study groups on energy conservation, energy efficiency, and renewable energy use (UN 1987). The Rio de Janeiro conference in 1992 marked the recognition of global warming as the key issue that would drive the development and use of energy resources globally (UN 1992). This changing perspective on the constraints on energy use and supply, over the past 40 years, raises some important questions, especially for less developed countries. This is the same time period when many countries of the global south have seen increasing economic growth and a development of their industrial capabilities. The recognition of global environmental constraints, concerns about local environmental pollution, and increasing difficulties in accessing finite fossil fuel reserves are proving to be significant challenges for these countries.

Evaluating these complex, and often conflicting demands on the energy system is the task of energy and economic models built to evaluate the linkages between key economic parameters and energy use. In the current context of climate change mitigation these models have been used to provide long-range forecasts of energy requirements and the consequent emissions in the future. A review of the range of such models, including both large macroeconomic models and technology-explicit energy models, built for India, is provided in this chapter. The review highlights some gaps in answering key questions, especially about the interlinkages between energy and economic and environmental variables. Some of these gaps are summarized here. Most models of the CGE and MARKAL variety estimate optimal solutions for single objective functions—either for utility maximization, surplus maximization, or energy cost minimization—assuming that a central planner or an efficient market decides monetary flow in the economy. There is an assumption that all agents will make rational choices and irrational producers and consumers will be driven out of the system. This hypothesis subsumes the assumption that agents have complete knowledge of all market parameters. These assumptions are significant departures from the real world and are likely to introduce significant uncertainties in model results even for more stable developed economies (Decanio 1979; Taylor and Taylor 2009). In addition to this, forecasts for economic parameters such as sectoral contributions to GDP in many models tend to not consider structural changes in the economy (Parikh and Parikh 2012; Parikh and Ghosh 2009; Shukla 2006). Forecasts for physical infrastructure, industrial production, and energy services are highly dependent on the sectors generating this demand and therefore in developing countries where a large part of the stock of infrastructure is yet to be built studies that do not consider the importance of the structural composition of the economy will tend to either under- or overestimate energy demands. Additionally the imposition of normative assumptions in the models built for less developed countries is often derived from the understanding of systems in the developed world and leads to either under-privileging some of the key and most immediate concerns of the less developed countries or a more limited and narrow use for the models. Finally, any one modeling method privileges one aspect of the problem over the other depending on the purpose for which it was originally meant. It is therefore not possible to understand the important energy-economy-emission linkages using only one model. This research aims to address some of these gaps, using a framework which integrates multiple modeling methodologies.

We propose here the use of an integrated modeling framework (IMF) where instead of an overarching objective function for welfare or utility maximization (as in the case of Computable General Equilibrium—CGE models) or a more limited optimization of energy costs relative to their demand at any given point of time (as in the MARKAL/TIMES variety of models), an integrated approach is used, where multiple modeling techniques are linked and used to address a broad array of questions related to the economy, energy systems, and the environment.

1.2   Scope of the Brief

A few key objectives were sought to be met as part of this work. Some of these objectives are discussed in this section, although the discussion in the brief will be broader and not restricted to these questions alone. The first objective that the work tries to address is to investigate whether the factors are likely to drive energy requirements and emissions in a developing economy, with a specific emphasis on estimating the impact of these largely macroeconomic factors on energy and emissions.

The second objective follows the first in attempting to evaluate how these requirements can be met under multiple constraints. The specific question of interest would be say, "what impact will environmental constraints have on fuel supply options in a capital scarce country?" The third objective is to them find ways to evaluate the potential impacts of certain energy and economic choices on economic growth, incomes, and equity.

For the specific case of India, the questions are further narrowed down to focus on particular aspects of India's energy and economic system. The role of economic structure, i.e., the sectoral composition of the economy, in determining the energy and emission intensity of GDP, is evaluated among other factors such as improvement in energy efficiency and change in the use of different fuel technologies. The options for energy supply under the multiple constraints of capital, emissions, and resources are also evaluated given the particular technical characteristics and constraints of the energy system in India. The impacts of high deployment of renewable energy sources, on other sectors of the Indian economy, on overall economic growth, and on incomes and equity are then assessed.

Various studies have attempted to address some of these questions using modeling techniques. A comprehensive overview of existing studies is provided in the next chapter along with a discussion on the strengths and weaknesses of these modeling approaches. To summarize briefly however, for the purpose of a more self-contained introduction to the text, it is observed that while each modeling methodology is useful for specific purposes, the arena of problems represented by the debate on climate change and development requires the addressing of multiple questions, often of equivalent importance, especially for developing countries. In this work therefore, it is proposed that to enable such a project, it is more useful to combine a range of modeling methodologies instead of relying on one method to deliver the answers. We do not claim that there is only one way in which to combine different methods. The choice of which models can and should be used together would very much depend on the set of questions that are being asked as well as the specific characteristics of the system and region being studied. We present here one possible combination of approaches that can provide a more comprehensive view of 3E linkages (energy-economy-environment) as an illustration of, and argument for, a different approach to 3E modeling.

The integrated modeling framework (IMF) proposed here combines three modeling approaches from the range of methods available, viz.—(i) decomposition analysis, (ii) optimization, and (iii) input-output analysis. The reason for choosing these three approaches is that they were found to be best suited for the analysis of the impact of environmental action (such as climate change mitigation interventions), and structural changes in the economy of a developing country such as India, on economic growth, incomes, income distribution, equity, and poverty alleviation. Figure 1.1 shows the schematic for the method used to integrate these three approaches.

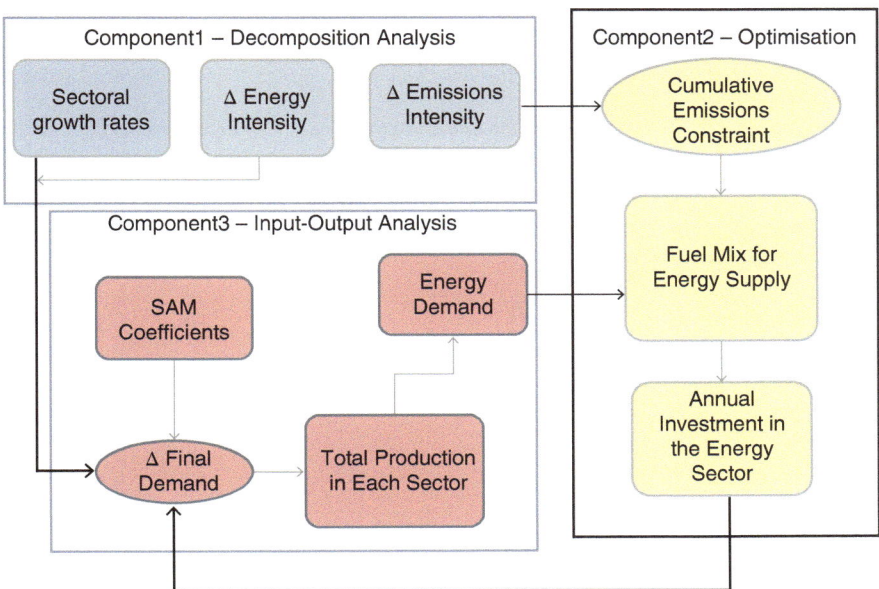

Fig. 1.1 Integrated modeling framework (IMF)—integrating three modeling approaches—decomposition analysis, I-O analysis, and constrained optimization

After a review of literature in Chap. 2, in Chaps. 3–5 each of the methodologies used in this analysis, viz., decomposition analysis, least cost optimization, and I-O analysis, respectively, is discussed with results for independent scenarios constructed and analyzed using these methods. Chapter 6 discusses the integration of the three components and presents an integrated modeling framework that can be used to answer key questions that are important for developing countries. The concluding chapter provides a summary of the work, advantages and contributions, limitations of the methods used, and directions for future work.

References

Brundtland, G., & Khalid, M. (1987). UN Brundtland commission report. Our common future.

DeCanio, S. J. (1979). Rational expectations and learning from experience. *The Quarterly Journal of Economics, 92*, 47–57.

Meadows, D. H., Meadows, D. H., Randers, J., & Behrens III, W. W. (1972). The limits to growth: A report to the club of Rome (1972).

Parikh, J., & Ghosh, P. P. (2009). Energy technology alternatives for India till 2030. *International Journal of Energy Sector Management, 3*(3), 233–250.

Parikh, J., & Parikh, K. (2012). Growing pains: Meeting India's energy needs in the face of limited fossil fuels. *IEEE Power and Energy Magazine, 10*(3), 59–66.

Shukla, P. R. (2006). India's GHG emission scenarios: Aligning development and stabilization paths. *Current Science, 90*(3), 384.

Smil, V. (2004). World history and energy. *Encyclopaedia of Energy, 6*, 549.

Taylor, L., & Taylor, L. (2009). *Reconstructing macroeconomics: Structuralist proposals and critiques of the mainstream*. Cambridge, MA: Harvard University Press.

UN General Assembly, United Nations Conference on the Human Environment, 15 December 1972, A/RES/2994, available at: https://www.refworld.org/docid/3b00f1c840.html [accessed 20 May 2019].

United Nations. (1972). *Declaration of the United Nations Conference on the Human Environment*. Retrieved December 5, 2015, from http://www.unep.org/Documents/Default.asp?DocumentID=97&ArticleID=1503.

United Nations Conference on Environment and Development, & Johnson, S. (1992). *The Earth Summit: The United Nations Conference on Environment and Development (UNCED)*. London: Graham & Trotman/Martinus Nijhoff.

Chapter 2
Review of Energy-Economy-Environment Models

Abstract Evaluating the complex demands on the energy system is the task of energy and economic models built to evaluate the linkages between key economic parameters and energy use. In the current context of climate change mitigation these models have been used to provide long-range forecasts of energy requirements and the consequent emissions in the future. Some of the aspects that these models try to address are (i) estimating energy consumption, considered a major area of concern; (ii) forecasting resource potential and reserves, mainly for oil and gas availability; (iii) the study of energy substitutions; and (iv) forecasting economic growth, income, and energy use and supply linkages for the future. Models that address these issues highlighted above range from simple bottom-up exercises undertaken to evaluate the economic viability of a certain fuel source to complex integrated energy planning models using multi-objective programming techniques linked with some forms of input-output and general equilibrium models. This chapter provides a review of the main trends and methodologies used in these models to analyze the issues outlined above. Some models constructed for the Indian energy and economic system are discussed in more detail. A section on the advantages as well as disadvantages of these models and their applicability to addressing the situation especially in developing economies such as India, as well as a brief introduction to the proposed integrated modeling framework, is also included in this chapter.

Keywords Energy forecasts · Input-output models · Computable general equilibrium models · Decomposition analysis

This chapter provides an overview of the debates in energy-economy-environment (3E) models by discussing their methodologies and assumptions. As the work presented here is illustrated for the case of India, the focus of the literature is also on modeling studies that have been done for India. However, studies done for other developing as well as developed countries have been considered here as well.

The first section of this chapter presents an overall review of energy models. The next section focuses specifically on the work done by others on analyzing historical energy and emission trends. The third section discusses the literature on input-output

Table 2.1 Classification of energy models by purpose

Purpose of model	Description	Important papers in the category
Estimating energy consumption	Forecast energy consumption and analyze sector-wise energy demand for a specific region over time	Lovins and Parisi (1977), Smil (1998), Lee (2005)
Forecasting resource potential and reserves	Studies mainly for oil and gas availability and timescales for which these will be available. Some early studies for coal as well	Hubbert (1975), Jevons (1906)
Studying energy substitutions	Replacement of one source of energy by another for a particular use where the objective is to forecast the demand for a particular fuel in the future	Marchetti (1977), Kemp (1994), Fouquet (2010)
Forecasting energy-economy linkages	Study linkages between economic growth, income, employment, and energy use and supply	Arbex and Perobelli (2010), Pollin and Chakraborty (2015)

modeling used for energy analysis. The last section discusses literature on technology-explicit energy modeling typically done using tools such as MARKAL or TIMES or similar other tools. Other studies that also propose integrated modeling methods are discussed in the final section of this chapter.

2.1 Overview of Energy Modeling Studies

Evaluating the complex demands on the energy system is the task of energy and economic models built to evaluate the linkages between key economic parameters and energy use. In the current context of climate change mitigation these models have been used to provide long-range forecasts of energy requirements and the consequent emissions in the future. Some of the broad aspects that these models try to address are briefly discussed here. These have been divided into four categories in terms of modeling approaches that try to estimate energy consumption or resource potential, try to study energy substitutions, or try to forecast energy demand and supply based on energy-economy linkages. These are shown in Table 2.1.

Models that address these issues highlighted above range from simple bottom-up exercises undertaken to evaluate the economic viability of a certain fuel source, say solar thermal or solar photovoltaic (Landsberg 1977), to complex integrated energy planning models using multi-objective programming techniques linked with some forms of input-output models (Hsu et al. 1988). A brief overview of the historical development of energy models is shown in Table 2.2.

Broadly, the modeling methodologies used to address the concerns outlined in Table 2.1, especially in the context of planning energy infrastructure for the future, can be divided into four categories—demand forecasting models, decompositions models, energy planning models, and economic models. This is a very broad

Table 2.2 Overview of energy models

Author	Model description
Stern (1977)	Quasi-equilibrium policy impact model for the supply of depletable resources with applications to crude oil
De Musgrove (1984)	MARKAL model to analyze minimum discounted cost configurations for the Australian energy system during the period 1980–2020
Macal et al. (1987)	An integrated supply and demand energy planning model for the state of Illinois
Hsu et al. (1988)	Integrated energy planning model using a multi-objective programming technique linked with the traditional Leontief input-output model
Rahman (1988)	Econometric energy-economy simulation model for energy policy studies
Bowe et al. (1990)	Markov model for engineering-economic planning
Joshi et al. (1992)	Simple linear decentralized energy planning model for a typical village in India
Badri (1992)	Model to analyze the demand for electricity in the residential, commercial, and industrial sectors of the United States
Suganthi and Jagadeesan (1992)	Mathematical programming energy-economy-environment (MPEEE) model. Maximizes the GNP/energy ratio based on environmental constraints
Hammond and Mackay (1993)	Forecasting model to project the oil and gas supply and demand to 2010 for the UK
Blakemore et al. (1994)	Econometric model used to study the effects of energy demand for the manufacturing sector (1970–1987) relating to the UK energy market
Reddy (1995)	A multi-logit model for fuel shifts in the domestic sector using the energy-ladder concept to study the effects of different factors on the selection of an energy carrier for cooking or water heating
Javeed Nizami and Al-Garni (1995)	A two-layered feed-forward artificial neural network forecasting model to relate the electric energy consumption to the weather data, global radiation, and population
Rao and Parikh (1996)	A trans-log econometric model based on time series data
Suganthi and Williams (2000)	An optimization model to determine the optimum allocation of renewable energy in various end uses in India for the period 2020–2021
Kumar et al. (2003)	Long-range energy alternative planning (LEAP) model for Vietnam
TERI (2006)	Linear programming MARKAL model over a 30-year modeling timeframe
Pal et al. (2015)	A nonlinear general equilibrium model for India
Parikh and Ghosh (2009)	Linear programming model, which uses the activity analysis framework to model the linkages between the national economy and environment

Source: Suganthi and Samuel (2012), Jebaraj and Iniyan (2006)

definition and made here only for the purpose of functional simplification. Techniques such as decomposition analysis (both index and structural) used to analyze past and existing trends in energy use and emissions and the impact of economic

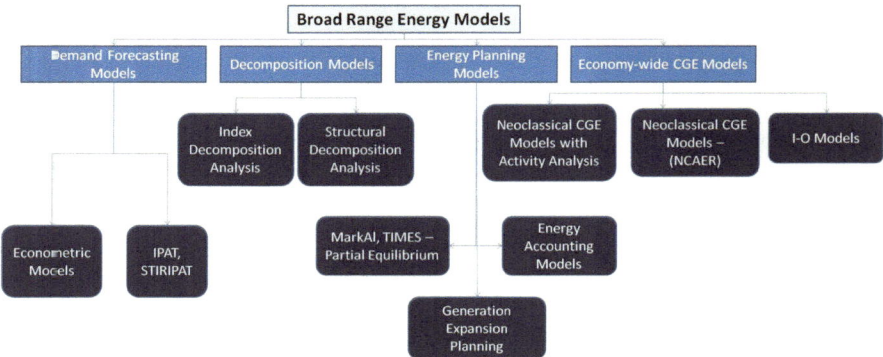

Fig. 2.1 Modeling methods and tools used to address key questions of energy use, supply, and its impact on economic output

activity on the same are classified as models here, as they aid in forecasting if required. However strictly speaking they are useful for the purpose of understanding existing and past trends of energy consumption and emissions. Figure 2.1 provides an overview of some of these modeling methodologies applied to study energy, economy, and environment linkages in India.

Methods for long-range forecasting of energy demand can broadly be classified into two categories—multivariate regression where key independent variables are regressed against the energy variable to obtain linear or nonlinear relationships that can then be used to forecast energy demand and emissions (TERI 2008). In the variety of models that use the IPAT or STIRPAT or related methods, energy or emission indicator is multiplicatively decomposed into two or three variables that have a causal relationship with the indicator (Hubacek et al. 2007). The IPAT equation is given below:

$$\text{Impact}(I) = \text{Population}(P) \times \text{Affluence}(A) \times \text{Technology}(T) \qquad (2.1)$$

STIRPAT is an extended form of the IPAT equation where STIR stands for stochastic impacts by regression (York et al. 2003). The relationship between the variables established through the identification of these drivers and their impact on the variable of interest at any given point of time is then extended or extrapolated to forecast impact for the future.

Index and structural decomposition methods evaluate past trends in certain energy or emission indicators by decomposing this indicator into its main driver variables. They provide a means to analyze the impacts of economic activity and the structural changes in economic activity on energy use and emissions. A study by Paul and Bhattacharya (2004) is one example of the use of this method for India.

The energy planning models include broadly three kinds modeling methodologies. Energy accounting models such as LEAP—long-range energy alternatives planning—provide a way to analyze energy flows from resource extraction to end use including losses. The model for "Long term Energy and Developmental Pathways for India" developed at the Indian Institute of Technology—Madras (IGCS 2014)—is one example of the application to India. Generation expansion planning models consist of optimization models applied to particular energy sectors or subsectors. They typically provide a way to plan for increasing energy supply for a given value of energy demand under multiple constraints on the sector, which can include technical, financial, and environmental constraints while optimizing costs (Chen et al. 2010).

Partial equilibrium models using tools such as MARKAL or TIMES to generate partial equilibrium in energy markets and compute scenarios that provide energy services at least cost under multiple technological constraints are the most widely used set of energy-environment models. Models by the Energy and Resources Institute for the Government of India (MoEF 2009) or the one of Shukla et al. (2008) use variations of this modeling method.

Economic models can also broadly be classified into three categories. Computable general equilibrium (CGE) models are the most common variety of models used for energy-economy analysis. The models use aggregate production functions computed based on an input-output framework and under assumptions of market equilibrium the production functions are used to forecast energy requirements for the future (Chadha 1998). Activity analysis models are extended CGE models that provide detailed analysis of particular activities or sectors (such as the energy sector) in the economy (Parikh et al. 2009). Structural path analysis, a method proposed by Defourny and Thorbecke, uses the I-O representation of national accounts to estimate how money travels through the economy and how a certain perturbation in the final demand could affect monetary flows and production in other sectors (Defourny and Thorbecke 1984). This third category of models is seldom used for energy analysis but can be very useful in exploring the impacts of, say, new investments on the economy as a whole and on specific sectors as well.

There is considerable overlap among these models in terms of the key variables they handle. For example, economic growth is a variable in most economic models and is used as an input in the energy models. The difference in the models lies in the degree of complexity of the models and their uses. Energy planning models are mostly cost feasibility models that are used to plan for energy in the existing time period. The energy supply-demand models are generally economy-wide static demand-supply analyses. The forecasting models mostly use regression analysis to forecast trends of energy use and supply for the future. Optimization models are usually economy-wide analyses of the energy sector with multiple constraints—cost, emissions, etc.

With increasing concerns of the impacts of climate change and the need to mitigate it, modeling exercises have started including all three parameters (energy, economy, and environment) in the analysis. Worldwide, the study of these linkages has been done by building macroeconomic or large-scale energy models. Many of these methodologies have also been used to model energy systems in India. There are large-scale energy models and also macroeconomic general equilibrium models that explore the linkage between energy and economy closely. The environmental impact is usually determined exogenously as an output of an energy-economy optimization exercise. A review of some of the major modeling methods applied in the context of India is given in the following sections.

2.2 Studies Examining Past Emissions and Emission Intensities

The greenhouse gas emissions in any country depend on a few key variables—the composition of fuels in primary energy production, the energy intensity in different sectors, the total economic activity, and the structural composition of the economy, i.e., the relative contribution of primary, secondary, and tertiary sectors to the economy. The total emissions in any country can then be expressed as a function of all these parameters as shown in Eq. (2.2) developed by Ang and Zhang (2000):

$$E = \sum_i I_i \times C_i \times S_i \times G \tag{2.2}$$

where E denotes the overall emissions in the economy; i denotes the sectors of the economy, e.g., $i = 1 =$ industry, $i = 2 =$ services, and $i = 3 =$ agriculture; I_i is the energy intensity of value added in the ith sector; C_i is the emission intensity of energy of the ith sector; S_i is the value added by sector i as a fraction of the total value added in the economy; and G is the total value added in the economy.

A decomposition of emissions using this primary equation can provide a method to evaluate the contribution of each factor to the total change in emissions. Such an analysis is important for developing countries as the structural composition of the economy in these countries is dynamic and subject to rapid change in the phase of economic transition. Galli (1998), Nag and Parikh (2000), Reddy and Ray (2010), Paul and Bhattacharya (2004), and Kojima and Bacon (2009) have analyzed past trends in energy and emission indicators for India for specific sectors. For example, Nag and Parikh (2000) discuss the emission intensity of the commercial sector and Reddy and Ray (2010) discuss the energy intensity of GDP in manufacturing. Paul and Bhattacharya (2004) discuss the decomposition analysis for emissions intensity in detail for four sectors (industry, services, agriculture, and transport) for a time span of 1980–1996. The discussions in Paul and Bhattacharya (2004) as well as in Kojima and Bacon (2009) focus on the application of decomposition methods to the Indian economy and the more recent changes in emission intensity, respectively. In the work presented as part of this brief we evaluate the trends in energy use and the changes in emissions between 1970 and 2008 for India, using decomposition indices, and also take this analysis forward by using the results from the decomposition analysis to construct scenarios for the future. Some of the results have been published in Kanitkar et al. (2015).

Studies for projecting future emissions such as the Report of the Low Carbon Committee (GoI (2011), Shukla (2006), Parikh and Gokarn (1993)) consider the potential changes in technology and efficiency while estimating future emissions within an overall economic growth rate. However, the evolution of the economic structure in the future is not discussed, making a business-as-usual economic trajectory an implicit assumption in the construction of future scenarios. In the work presented in this brief, we estimate future emissions including those for scenarios where the structural composition of the economy changes.

2.3 Optimal Energy Pathways

Most of the literature that discusses optimum energy pathways for countries is based on assumptions of some "business-as-usual" trajectory which is constructed as the representation of a "future" that will be based on an extrapolation of past trends. Among the problems associated with such approaches is the inherent ambiguity in defining the business-as-usual trajectory that may typically contain a number of assumptions regarding efficiency improvements, structural composition of the economy, and improvement in the productivity of factors of production. Such models, that also make predictions based on extrapolating from the current structure of the economy (Nordhaus 2008; Manne and Richels 2005) or the extrapolation of energy demands alone (as in bottom-up energy demand models) (Dean et al. 1992; Rai and Victor 2009; Mallah and Bansal 2010), are likely to significantly underestimate the future requirements of physical infrastructure, industrial production, and energy services for developing countries. More recently modeling studies have been undertaken to determine the impact of cumulative emissions targets on various sectors of the economy (Anderson et al. 2008; Wang and Watson 2010). Such a study however has not been undertaken for India. Also, the studies cited above do not provide concrete estimates of the additional financial burden that may accrue due to increasingly stringent emission targets.

Detailed sector-level analyses have been done for many countries to determine the quantum of changes that would be required in order to address the question of climate change mitigation. Bottom-up energy models (very often built using the MARKAL or TIMES platforms) also typically extrapolate historical trajectories of economic growth and their consequent implications for energy demands (TERI 2006). Total emissions are usually an output from such models. The TERI MARKAL Model is the most detailed disaggregated energy model for India and is therefore explained a little more in detail in the following subsection.

2.3.1   The TERI MARKAL Model

MARKAL (market allocation) is a generic model tailored by the input data to represent the evolution over a period of usually 30–50 years of a specific energy system at the global, national, regional, or state level. It is a bottom-up cost-minimization, linear programming model for the energy sector with a potential to internalize environmental considerations and study the effects thereof. The TIMES modeling framework is an improved version of the MARKAL framework for energy modeling. The only major difference between the two frameworks is that while in the MARKAL modeling framework energy demand is considered to be an exogenously supplied parameter, in the TIMES framework the energy demand is elastic to its own price. The objective function is the minimization of the total energy system cost, time discounted over the planning horizon.

TERI's MARKAL model is the application of the general MARKAL model to the specific case of India. In the TERI model, each year, the total cost includes the following elements—(i) annualized investment costs of energy technologies, (ii) fixed and variable annual operation and maintenance costs (O&M) of technologies, (iii) costs of exogenous energy and material imports and national resource production (e.g., mining), (iv) revenue from exogenous energy and material exports, (v) fuel and material delivery costs, and (vi) taxes and subsidies associated with energy sources, technologies, and emissions. The objective function is specified as follows:

$$\text{Minimize TDSC} = \sum_{t=1}^{t=\text{NPER}} (1+d)^{\text{NYRS}(1-t)} \times \text{Anncost}(r,t)$$
$$\times \left(1 + (1-d)^{-1} + (1+d)^{1-\text{NYRS}} \right) \qquad (2.3)$$

where TDSC is the total discounted system cost, $\text{Anncost}_{r,t}$ is the annual cost for period t, Anncost = Import cost − Exports Revenue − Salvage value − Emission fees, NPER is the number of periods in the planning horizon, NYRS is the number of years in each period t, and d is the discount rate for each period.

The Indian energy sector is disaggregated into five major energy-consuming sectors, namely, agriculture, commercial, industry, residential, and transport sectors. Each of these sectors is further disaggregated to reflect the sectoral end-use demands. For example, the industrial sector is disaggregated into eight energy-consuming industries, viz., chlor-alkali (soda ash, and caustic soda), aluminum, iron and steel, cement, textile, fertilizer, pulp and paper, and other manufacturing units grouped as other industries. Similarly in the residential sector, the demand is projected for lighting, space conditioning, cooking and refrigeration, etc., separately for urban and rural households to account for the differences in lifestyles and choice of fuel and technology options. The end-use demands in each subsector are exogenously provided based on the outputs of other CGE models. Energy demand is driven by population and GDP growth.

It is assumed that urban population will increase from 29% in 2006 to 42% in 2030. With high GDP growth rates, BPL houses are predicted to reduce and high expenditure households expected to increase. A GDP growth rate of 6.7% (CAGR) is expected to lead to an almost complete elimination of the "most poor" category (expenditure of less than Rs. 615/month/household) in the model. An increase in GDP is considered as a proxy for increase in purchasing power which will lead to enhanced education and health services.

On the supply side, the model considers various energy resources that are available both within the country and abroad for meeting various end-use demands. These include both the conventional energy sources such as coal, oil, natural gas, hydro, and nuclear and the renewable energy sources such as wind, solar, and biomass. The availability of each of these fuels is restricted based on the expected growth in national production and plans of the Government of India. Further, various conversion and process technologies characterized by their respective investment costs, operating and maintenance costs, technical efficiency, life, etc., to meet the sectoral end-use demands, are also incorporated in the model.

The outputs of the TERI model include—(i) a set of investments in all technologies selected by the model at each period; (ii) a set of operating level of all technologies at each period; (iii) the quantities of each fuel produced, imported, and/or exported at each period; (iv) sectoral energy consumption (aggregate), fuel mix, and emissions at each period; (v) the emissions of greenhouse gases (GHG) and pollutants at each period; and (vi) the overall energy system's discounted cost.

The TERI-Energy Roadmap which discusses this model and its assumptions in detail talks about how a decline in the contribution of agriculture to GDP has not been accompanied by a corresponding decline in the number of people engaged in agriculture (which is still 60% according to Census 2001 data). While it projects these same trends of the past into the future, it does not discuss the implications of this structure of the economy (with the service sector dominating growth) on other parameters such as incomes or income distributions. However, the advantage of this model is this detailed explanation of the demand side, which can factor in efficiency improvement, technology saturation, and other important parameters.

2.4 Energy Modeling Using Input-Output Methods

The work done on using the input-output methods to analyze the impacts of energy policy on incomes and equity as part of this research uses an extended input-output matrix. The social accounting matrix (SAM) is a closed input-output matrix with respect to households. It is a particular representation of the macroeconomic accounts of a socioeconomic system, which captures the transactions and transfers between all economic agents in the system (Pyatt and Round 1985; Reinert and Roland-Holst 1997). It records transactions taking place during an accounting period, usually taken to be 1 year, and the economic flows are recorded in the form of a square matrix. One of the important features of a SAM is that households and household groups are the most important institutions in the framework. The matrix can only be called a "social" accounting matrix if it represents some distributional features of the household sector (Round 2003).

The method of using social accounting matrices began with the work of Sir Richard Stone in the 1960s based on data for the UK and some other industrialized countries. These ideas were further developed and used to help study issues such as poverty and income distribution in developing countries by Pyatt, Thorbecke, and others in the 1970s (Pyatt et al. 1976). Some of the earliest SAM-based multiplier studies have been for Sri Lanka (Pyatt and Round 1979), Botswana (Hayden and Round 1982), Korea (Defourny and Thorbecke 1984), Indonesia (Thorbecke 1992), and more recently Ghana (Powell and Round 2000) and Vietnam (Tarp et al. 2002). All these studies aim to examine the nature of multiplier effects of an income injection in one part of an economic system on the incomes of socioeconomic groups of households.

The SAM itself is not a model. It is a representation of a set of macro data for an economy. The basic approach to SAM-based multiplier models is to compute column shares (column coefficients) from a SAM in order to compute matrix multipliers. In doing so, one or more of the accounts must be designated as being exogenous. Usually, the transactions in the government account, the capital account, and the rest of the world account are regarded as exogenous (Pyatt and Round 1985).

In India, economists argue that construction of a SAM has been difficult because of data constraints, particularly for income distributions (Saluja and Yadav 2006). Sarkar and Subbarao (1981) built a SAM for India where the private income of the economy was divided into three parts, viz., agricultural wage income, nonagricultural wage income, and nonagricultural nonwage income. The SAM for 1977–1978 constructed by De Janvry and Subbarao (1986) grouped households into seven social classes characterized by sources of income, viz., (1) rural landless agricultural workers, (2) rural small peasants, (3) rural middle farmers, (4) rural large peasants, (5) urban workers, (6) urban marginals, and (7) urban capitalists. A detailed SAM was constructed for the Indian economy for 1994–1995 for 60 sectors of the economy with two factors of production and six categories of occupational households, separately, for rural and urban areas by Pradhan et al. (1999) which was later updated to 1997–1998 and to 2002–2003 (Pradhan et al. 2006). One of the SAMs used in this analysis is constructed by Saluja and Yadav (2006) for 2003–2004 and consists of 73 production sectors, 2 factors of production, and 5 household classes by expenditure levels separately for rural and urban areas. A SAM has also been constructed by Pradhan et al. for the year 2007–2008. This SAM is similar to the SAM for 2003–2004 in its design but for one major difference in that the household classification in this SAM is not based on expenditure but on occupational categories. The 2007–2008 SAM is more recent and therefore using it may provide results that are more accurate. However, there were two disadvantages in using this SAM. The classification of household classes in the 2007–2008 SAM is based on occupations and not income (or expenditure) classes. This makes it difficult to actually evaluate any measure of equality as there is a wide income range within every occupational category. The results for income distribution are thus skewed—inequality is undercalculated (Pieters 2010). The other disadvantage is that the results obtained from using the SAM 2007–2008 cannot be validated against some other studies as most other studies for energy and emission projections for the future have used the SAM 2003–2004.

Other studies have used the social accounting framework to construct energy scenarios for the future. However, most recent studies estimate a CGE model using data from an input-output table and use an overall objective function (that maximizes utility) to determine the outcome by assuming economy-wide equilibrium, viz., Parikh and Ghosh (2009) and Pradhan and Ghosh (2010). There are broadly two categories within the CGE models. The more commonly constructed neoclassical CGE models begin with the hypothesis of optimizing agents and full employment. The incorporation of social classes is usually a secondary feature of these models. On the other hand there are structuralist CGE models which consider that structural characteristics of the economy are fundamental to its behavior (Taylor 1990). The work presented in this brief does not use a CGE model but develops a forecasting model using the social accounting matrix and uses structural path analysis based on decomposition multipliers calculated for the SAM to determine how a particular investment travels through the economy and affects different sectors. The model is similar to a simple Leontief forecasting model, but done using a SAM. There is no optimization in the model, so assumptions about rational agents taking decisions regarding production and consumption in the economy can be avoided. The model instead runs a set of simulations for different scenarios based on certain assumptions of structural linkages and behavior of exogenous factors—government expenditure and capital investment in this case. The results from the model are compared to other models that have also evaluated energy and economic variables for India. It is therefore necessary to first give a brief review of the main models that have been built for India, which is provided in the next section.

In 2010, the Ministry of Environment and Forests (now the Ministry of Environment, Forests and Climate Change) had commissioned four studies, of which two were top-down macroeconomic models that used the SAM for 2003–2004 as the basis for defining inter-sectoral linkages. The methodologies used in these two models are briefly described here.

2.4.1 NCAER Model

The model built by NCAER (National Centre for Applied Economic Research) is a top-down model, based on a neoclassical CGE framework that includes institutional features peculiar to the Indian economy. The economy is represented by 37 sectors in this nonlinear model. It includes the interactions of producers, households, government, and the rest of the world in response to relative prices, given certain initial conditions and exogenously given set of parameters. Producers act as profit maximizers in perfectly competitive markets; that is, they take factor and output prices (inclusive of any taxes) as given and generate demands for factors so as to minimize unit costs of output. Consumers maximize utility subject to their budget constraints. Also households save and pay taxes to the government. Furthermore, households are classified into five rural and four urban socioeconomic groups. The primary energy sectors used are coal, oil, gas, hydro, nuclear, and biomass; it is possible to include supply constraints for each energy form. GHG emissions arise in fixed coefficients for each energy form and for specified industrial processes. The factors of production include energy inputs and "primary inputs"—capital, land, and different types of labor. Technological change is described in terms of total factor productivity growth (TFPG) and autonomous energy efficiency index (AEEI). These are exogenous model inputs.

For the illustrative scenario—which is the scenario used by the MoEF (Ministry of Environment and Forests)—it is assumed that the TFPG is 3% p.a. based on "available empirical evidence," "expert opinion," and "reasoned judgment on different baseline GDP scenarios." Improvement in specific energy consumption is incorporated in the model by using the values for the autonomous energy efficiency improvement (AEEI) used in other carbon emission abatement models (EPPA for example). AEEI occurs in all sectors except the primary energy sectors (coal, crude petroleum, and natural gas) and the refined oil sector. The total AEEI is assumed to be 1.5% on the basis of an "encouraging improvement in energy efficiency in the last two decades" and "being a typical value often used by other modelers." These values are assumed to remain constant over the entire model period.

Model outputs include GDP (and GDP growth), prices, incomes, quantities of imports and exports, final consumption, and government demands, besides energy use and GHG emissions. The model is predictive and simulates the effects of particular policy and parameter assumptions. The model has been calibrated to the benchmark "equilibrium" data set of the Indian economy for the year 2003–2004 as given by the social accounting matrix (SAM) available for 2003–2004.

2.4.2 IRADe Activity Analysis Model

The IRADe model is a linear programming model based on the input-output framework. The model uses the activity analysis framework to model linkages between the national economy and environment. It maximizes an objective function, which is the discounted sum of total consumption streams given the resources available to economy and the various technological possibilities for using them. The model has endogenous income distribution. It traces well-being effects for the low-income groups by examining the incidence of absolute poverty in the population. The differences in consumption patterns among different income classes are represented using a linear expenditure system (LES) of equations for each consumer class. The model has an endogenous income distribution change which impacts the structure of consumption demand in the economy, as population in a lower income

Table 2.3 Resource constraints for the IRADe activity analysis model

Coal	Specified to grow by at the most eight times the base year output
Natural gas	Specified to grow by at the most three times the base year output
Crude oil	Specified to grow by at the most two times the base year output
Wood gasification	At max 2% of forestry output can be used for power generation
Hydro	Maximum output of power from hydro is 440 billion kWh
Wind	Maximum output from wind power is 175 billion kWh
Nuclear	Maximum output from nuclear power is 375 billion kWh (optimistic IEP scenario)
Natural gas	At max 40% of domestic availability can be used for power generation

Source: Parikh et al. (2009)

class today will move to a higher income class in the future as income growth takes place. There are terminal conditions on stock variables such as natural resources in the model, which the authors call "sustainability constraints" given in Table 2.3.

The structure of the input-output matrix which is used in the model is linked to the growth model on the one hand and an analysis of the energy sector on the other hand. The stated objective of the IRADe model is "to explore various alternatives to meet energy demand and their impact on the macro economic variables in general and poverty in particular." The authors estimate a CGE model for India using the SAM for 2003–2004. The model allows for a two-way interaction between energy sectors and other sectors of the economy. The model maximizes the present discounted value of private consumption over the planning period, which is 2003–2030, subject to demand and supply constraints using Eq. (2.4):

$$\text{Max } U = \sum_{t=0}^{T} \frac{\text{POP}_t \times \text{PC}_t}{(1+r)^t} + \text{P}'\text{C} \tag{2.4}$$

where POP_t is the population at time t, PC_t is the per capita consumption at time t, T is the time horizon under consideration, r is the discount rate used in the model, and PC is the average per capita consumption beyond the time period under consideration.

The model has each commodity being produced by one production activity, except electricity. To produce power the model employs renewable (wind, solar thermal, solar photo voltaic, wood gasification) and nuclear based technologies apart from the traditional technologies of thermal, hydro, and gas. Coal, crude, natural gas, and electricity are energy inputs into the model. The model ensures commodity-wise equilibrium between demand and supply as shown in Eq. (2.5):

$$C_{it} + G_{it} + I_{it} + \text{IO}_{it} + E_{it} \leq Y_{it} + M_{it} \tag{2.5}$$

where C is the private consumption of households of commodity i at time t, G is the government consumption of commodity i at time t, I is the demand for investment

Table 2.4 Annual reduction in input use between 2003 and 2030 (AEEI)

Coal	0.77%
Crude oil and natural gas	1.15%
Petroleum products	0.96%
Electricity	0.38%

goods in the economy, IO is the intermediate consumption of goods at time t, E is the amount of export of commodity i at time t, Y is the national production of commodity i at time t, and M is the imports of commodity i at time t.

The key endogenous variables in the model are the intermediate demand for goods, income distribution of households, and trade. The key exogenous variable is government consumption. Limits, constraints, and control parameters are exogenously specified in the model. The key control parameters are (i) the autonomous energy efficiency improvement (AEEI) given in Table 2.4; (ii) the total factor productivity growth (TFPG), the value for which has not been specified in the paper, but a value of 1.5% improvement per annum is typically used; (iii) the incremental capital output ratio (ICOR) which is the ratio of investment to growth. The higher the ICOR, the lower the productivity of capital. The ICOR can be thought of as a measure of the inefficiency with which capital is used; (iv) the maximum marginal savings rate, which is specified to be 40% in the model; (v) the government consumption is specified to grow at 9% p.a. in the model; (vi) a monotonicity constraint is imposed on the growth rate of private consumption—a minimum growth rate of 6% p.a. is imposed on consumption of households and firms; (vii) the model uses a social discount rate of 10%.

The model assumes an income parity of 2.34 between rural and urban areas. This means that rural and urban income and therefore their consumption are allowed to grow in this ratio only. A major assumption in the model is that for demand-side management. It is assumed that an amount of Rs. 4 is required to save Re. 1 worth of electricity requirement per year. The total investments required in the energy sector vary significantly with this assumption. CGE models use an overall objective function to determine the outcome by assuming that the coefficient values calculated for the base year represent equilibrium, viz., Parikh and Ghosh (2009) and Pradhan and Ghosh (2012). A notable exception to the CGE variety of models is the study to determine the impacts of energy policy on incomes and other distributional factors for Indonesia by Hartono et al. (2008).

2.5 Summary and Gaps

A review of these existing models highlights their specific relevance in advancing our understanding of the energy-economy-environment (3E) linkages as well as certain gaps in addressing the key questions that are relevant at this juncture. Some of these gaps are summarized here.

1. Model outcomes, especially in the case of computable general equilibrium (CGE) models, are based on values of key variables, many of which either cannot be anticipated or their probabilities cannot be placed within acceptable bounds so as to restrict the range of potential outcomes (Weinberg 1979). The use of these models for long-range forecasting therefore introduces significant uncertainty in the results. For example many models use economic growth as an input parameter into the model, based on which energy requirements are then calculated with the help of coefficients estimated for this purpose (TERI 2008; Parikh and Ghosh 2009). However, the value of economic growth itself depends on many assumptions, e.g., the structure of the economy or the stage of economic development, even if the other more fuzzy variables such as the regional and global political climate within which the economic growth manifests are ignored.

2. Most models especially belonging to the CGE variety estimate optimal solutions for single objective functions—for either utility maximization, surplus maximization, or energy cost minimization—assuming that a central planner or an efficient market decides monetary flow in the economy. This can hide the reality of the many distortions that actually exist in economic systems. Also, the key assumption in this framework of modeling is that each agent in the economic system is a rational agent with full prior knowledge of market parameters. This assumption is central even to the MARKAL-TIMES variety of models. The optimum solution is reached when the rational self-interest of each agent is maximized. However, key parameters in the model are very often exogenously fixed to arrive at such an optimal solution (Taylor and Taylor 2009). For example, in the CGE model built for the Indian economy by the National Centre for Applied Economic Research, 18 parameters are fixed exogenously in order to arrive at an optimum solution (MoEF 2009). This is likely to introduce some significant uncertainties in the models.

3. Models typically based on assumptions of equilibrium between demand and supply for a base year tend to extrapolate existing economic trends into the future and do not account for structural change in the economy. This is true even for multivariate regression models that use existing trends in drivers to forecast values for the future. Such models are likely to underestimate the future requirements of physical infrastructure, industrial production, and energy services for developing countries and the constraints that would be progressively placed upon the same, especially given that a large part of the stock of infrastructure that will be required in developing countries is yet to be built. Also in these models, sectoral changes take place under the objective of utility maximization based on some broad overall constraints on consumption and aggregated production functions. It is difficult to isolate and study the impact of a certain decision on a specific industry or income class. This also makes it difficult to construct explicit policy scenarios for poverty eradication or equitable growth.

4. Long-range energy demand forecasting, while necessary in energy planning exercises, is in itself a source of uncertainty in many models. While this cannot be completely managed within the modeling framework itself, it is necessary to highlight the wide range of issues that are associated with attempts to forecast

energy requirements. For developing countries, one major problem with demand forecasting is the potential changes in economic structures which have been discussed before. The other problem is one of technological development and diffusion. The example of demand-side management (DSM) illustrates this point effectively. Most of the literature on DSM in developing countries points to the need for more effective institutional arrangements for such programs to succeed. For example, Harish and Kumar (2014) and Vashishtha and Ramachandran (2006) point to the need for effective regulatory and institutional measures in the energy sector in India. There are also other peculiar characteristics that need to be considered while assessing the effects of any such program. For example a number of authors have pointed out the significant levels of rebound that exists for energy efficiency solutions across different sectors in India (Chakravarty et al. 2013; Roy 2000). There are techniques that are being developed to address some of these concerns to a certain extent now (Ardakani and Ardehali 2014); however while these may be useful at the level of sectoral demand forecasting, for macroeconomic energy-environment analysis addressing the efficacy of demand forecasts remains a challenge.

5. It is difficult to understand the important 3E linkages using only one modeling methodology as each method privileges some aspect of the problem. Technology-explicit energy models like MARKAL and TIMES are built specifically to model the technological aspects of the energy sector and cannot be used to address questions about the economy in general and about incomes and income distributions within the economy. On the other hand CGE models consider the economy as a whole at the macroeconomic level using aggregated production functions and therefore can neither capture the details of the energy sector, nor capture the details of the specific intermediate and end-use sectors of the economy.

2.6   Developing an Approach to an Integrated Modeling Framework (IMF)

In this work therefore, an alternative approach is suggested to address some of these concerns. We propose the use of an integrated modeling framework (IMF) where instead of an overarching objective function for welfare or utility maximization (as in the case of computable general equilibrium—CGE models) or a more limited optimization of energy costs relative to their demand at any given point of time (as in the MARKAL/TIMES variety of models), an integrated approach is used, where multiple modeling techniques are linked and used to address a broad array of questions related to the economy, energy systems, and the environment.

The IMF includes three main sub-models: (i) The first sub-model is a simple index decomposition analysis of the past trend in emissions intensity of GDP into its parametric components—primarily energy intensity of GDP and structural composition of the economy. This component provides the basis on which scenarios are constructed in the next two components of the IMF. (ii) The second sub-model is used to evaluate the optimum energy supply pathways under various resource and emission constraints, using simple linear programming techniques. The objective of this component is to develop a methodology for determining energy options, specifically for the power sector, under a variety of scenarios that capture the technical and economic characteristics of the power sector. (iii) The third sub-model is an input-output model that is used to evaluate the impact of energy policy on income and equity. The main objective in this part of the work is to build a methodology through which trade-offs associated with policy decisions in the power sector can be understood and quantified. This framework is demonstrated in this work for the case of India.

References

Anderson, K., Bows, A., & Mander, S. (2008). From long-term targets to cumulative emission pathways: Reframing UK climate policy. *Energy Policy, 36*(10), 3714–3722.

Ang, B. W., & Zhang, F. Q. (2000). A survey of index decomposition analysis in energy and environmental studies. *Energy, 25*(12), 1149–1176.

Arbex, M., & Perobelli, F. (2010). Solow meets Leontief: Economic growth and energy consumption. *Energy Economics, 32*(2010), 43–53.

Ardakani, F. J., & Ardehali, M. M. (2014). Novel effects of demand side management data on accuracy of electrical energy consumption modeling and long-term forecasting. *Energy Conversion and Management, 78*, 745–752.

Badri, M. A. (1992). Analysis of demand for electricity in the United States. *Energy, 17*(7), 725–733.

Blakemore, F. B., Davies, C., & Isaac, J. G. (1994). UK energy market: An analysis of energy demands. Part I: A disaggregated sectorial approach. *Applied Energy, 48*(3), 261–277.

Bowe, T. R., Dapkus, W. D., & Patton, J. B. (1990). 5.3. Markov models. *Energy, 15*(7–8), 661–676.

Chakravarty, D., Dasgupta, S., & Roy, J. (2013). Rebound effect: How much to worry? *Current Opinion in Environmental Sustainability, 5*(2), 216–228.

Chadha, R. (1998). *The impact of trade and domestic policy reforms in India: A CGE modeling approach.* Ann Arbor: University of Michigan Press.

Chen, Q., Kang, C., Xia, Q., & Zhong, J. (2010). Power generation expansion planning model towards low-carbon economy and its application in China. *IEEE Transactions on Power Delivery, 25*(2), 1117–1125.

Dean, A., Hoeller, P., & Organisation for Economic Co-operation and Development. Economics Dept. (1992). *Costs of reducing CO_2 emissions: Evidence from six global models (Vol 2).* Paris: OECD.

Defourny, J., & Thorbecke, E. (1984). Structural path analysis and multiplier decomposition within a social accounting matrix framework. *The Econometrics Journal, 94*(373), 111–136.

De Musgrove, A. R. (1984). A linear programming analysis of liquid-fuel production and use options for Australia. *Energy, 9*, 281–302.

Fouquet, R. (2010). The slow search for solutions: Lessons from historical energy transitions by sector and service. *Energy Policy, 38*(11), 6586–6596.

Galli, R. (1998). The relationship between energy intensity and income levels: forecasting long term energy demand in Asian emerging countries. *The Energy Journal, 19*, 85–105.

GoI. (2011). Planning Commissions, Government of India. Interim Report of the Expert Group on Low Carbon Strategies and Inclusive Growth. Retrieved December 6, 2015, from http://planningcommission.nic.in/reports/genrep/Inter_Exp.pdf

Hammond, G. P., & Mackay, R. M. (1993). Projection of UK oil and gas supply and demand to 2010. *Applied Energy, 44*, 93–112.

Harish, V. S. K. V., & Kumar, A. (2014). Demand side management in India: Action plan, policies and regulations. *Renewable and Sustainable Energy Reviews, 33*, 613–624.

Hartono, D., & Resosudarmo, B. P. (2008). The economy-wide impact of controlling energy consumption in Indonesia: An analysis using a Social Accounting Matrix framework. *Energy Policy, 36*(4), 1404–1419.

Hayden, C., & Round, J. I. (1982). Developments in social accounting methods as applied to the analysis of income distribution and employment issues. *World Development, 10*(6), 451–465.

Hsu, G. J., Leung, P., & Ching, C. T. (1988). Energy planning in Taiwan: An alternative approach using a multi-objective programming and input-output model. *The Energy Journal, 9*(1), 53–72.

Hubbert, M. K. (1975). Survey of world energy resources. *Energy Sources Future, 1*, 3–38.

Hubacek, K., Guan, D., & Barua, A. (2007). Changing lifestyles and consumption patterns in developing countries: A scenario analysis for China and India. *Futures, 39*(9), 1084–1096.

Indo-German Centre for Sustainability. (2014). *Long term energy and developmental pathways for India.* Chennai: IIT Madras. Retrieved April 28, 2016, from http://www.worldenergyoutlook.org/weomodel/investmentcosts/.

Janvry, A. D., & Subbarao, K. (1986). Agricultural price policy and income distribution in India. In *Studies in economic development and planning* (Vol. 43). Oxford: Oxford University Press.

Javeed Nizami, S. S. A. K., & Al-Garni, A. G. (1995). Forecasting electric energy consumption using neural networks. *Energy Policy, 23*, 1097–1104.

Jebaraj, S., & Iniyan, S. (2006). A review of energy models. *Renewable and Sustainable Energy Reviews, 10*(4), 281–311.

Jevors, W. S. (1906). *The coal question: An inquiry concerning the progress of the nation, and the probable exhaustion of our coal-mines.* London: Macmillan.

Joshi. B., Bhatti, T. S., & Bansal, N. K. (1992). Decentralized energy planning model for a typical village in India. *Energy, 17*(9), 869–876.

Kanitkar, T., Banerjee, R., & Jayaraman, T. (2015). Impact of economic structure on mitigation targets for developing countries. *Energy for Sustainable Development, 26*, 56–61.

Kemp, R. (1994). Technology and the transition to environmental sustainability: The problem of technological regime shifts. *Futures, 26*(10), 1023–1046.

Kojima, M., & Bacon, R. (2009). *Changes in CO_2 emissions from energy use: A multicountry decomposition analysis.* Washington: World Bank.

Kumar, A., Bhattacharya, S. C., & Pham, H. L. (2003). Greenhouse gas mitigation potential of biomass energy technologies in Vietnam using the long range energy alternative planning system model. *Energy, 28*(7), 627–654.

Landsberg, P. T. (1977). A simple model for solar energy economics in the UK. *Energy, 2*(2), 149–159.

Lee, C.-C. (2005). Energy consumption and GDP in developing countries: a cointegrated panel analysis. *Energy Economics, 27*(2005), 415–427.

Lovirs, A. B., & Parisi, A. J. (1977). *Energy strategy: The road not taken?* Collingwood: Friends of the Earth Australia.

Maca, C. M., Bragen, M. J., & Marshall, J. E. (1987). An integrated energy planning model for Illinois. *Energy, 12*(12), 1239–1250.

Manne, A. S., & Richels, R. G. (2005). MERGE: An integrated assessment model for global climate change. In *Energy and Environment* (pp. 175–189). New York: Springer.

Mallah, S., & Bansal, N. K. (2010). Allocation of energy resources for power generation in India: Business as usual and energy efficiency. *Energy Policy, 38*(2), 1059–1066.

Marchetti, C. (1977). Primary energy substitution models: On the interaction between energy and society. *Technological Forecasting and Social Change, 10*(4), 345–356.

MoEF. (2009). India's GHG emissions Profile, Results of five climate modeling studies.

Nag, B., & Parikh, J. (2000). Indicators of carbon emission intensity from commercial energy use in India. *Energy Economics, 22*(4), 441–461.

Nordhaus, W. D. (2008). *A question of balance: Weighing the options on global warming policies.* New Haven: Yale University Press.

Pal, B. D., Ojha, V. P., Pohit, S., & Roy, J. (2015). An environmental computable general equilibrium (CGE) model for India. In *GHG emissions and economic growth* (pp. 73–93). New Delhi: Springer.

Parikh, J., & Ghosh, P. P. (2009). Energy technology alternatives for India till 2030. *International Journal of Energy Sector Management, 3*(3), 233–250.

Parikh, J., Panda, M., Ganesh-Kumar, A., & Singh, V. (2009). CO$_2$ emissions structure of Indian economy. *Energy, 34*(8), 1024–1031.

Parikh, J., & Gokarn, S. (1993). Climate change and India's energy policy options: New perspectives on sectoral CO$_2$ emissions and incremental costs. *Global Environmental Change, 3*(3), 276–291.

Paul, S., & Bhattacharya, R. N. (2004). CO$_2$ emission from energy use in India: A decomposition analysis. *Energy Policy, 32*(5), 585–593.

Pradhan, B. K., & Ghosh, J. (2012). *The impact of carbon taxes on growth emissions and welfare in India: A CGE analysis.* New Delhi: Institute of Economic Growth, University of Delhi.

Pradhan, B. K., Sahoo, A., & Saluja, M. R. (1999). A social accounting matrix for India, 1994–95. *Economic and Political Weekly, 34*, 3378–3394.

Pradhan, B. K., Saluja, M. R., Singh, S. K., & Singh, S. K. (2006). *Social accounting matrix for India: Concepts, construction and applications.* New Delhi: Sage.

Pieters, J. (2010). Growth and inequality in India: Analysis of an extended social accounting matrix. *World Development, 38*(3), 270–281.

Pollin, R., & Chakraborty, S. (2015). An Egalitarian Green Growth Program for India. *Economic and Political Weekly, 42*(2015), 38–52.

Powell, M., & Round, J. I. (2000). Structure and linkage in the economy of Ghana: A SAM approach. In *Economic reforms in Ghana: Miracle or mirage* (pp. 68–87). Borough of Melton: James Currey.

Pyatt, G., Thorbecke, E., & Emmerij, L. (1976). *Planning techniques for a better future: A summary of a research project on planning for growth, redistribution and employment.* Geneva: International Labour Office.

Pyatt, G., & Round, J. I. (1979). Accounting and fixed price multipliers in a social accounting matrix framework. *The Economic Journal, 89*(356), 850–873.

Pyatt, G., & Round, J. I. (1985). *Social accounting matrices: A basis for planning.* Washington: The World Bank.

Rahman, S. H. (1988). Aggregate energy demand projections for India: an econometric approach. *Pacific and Asian Journal of Energy, 2*, 32–46.

Rai, V., & Victor, D. (2009). Climate change and the energy challenge: A pragmatic approach for India. *Economic and Political Weekly, 44*(31), 78–85.

Rao, R. D., & Parikh, J. K. (1996). Forecast and analysis of demand for petroleum products in India. *Energy Policy, 24*(6), 583–592.

Reddy, B. S. (1995). A multilogit model for fuel shifts in the domestic sector. *Energy, 20*(9), 929–936.

Reddy, B. S., & Ray, B. K. (2010). Decomposition of energy consumption and energy intensity in Indian manufacturing industries. *Energy for Sustainable Development, 14*(1), 35–47.

Reinert, K. A., & Roland-Holst, D. W. (1997). Social accounting matrices. In *Applied methods for trade policy analysis: A handbook* (pp. 94–121). Cambridge: Cambridge University Press.

Round, J. (2003). Social accounting matrices and SAM-based multiplier analysis. In *The impact of economic policies on poverty and income distribution: Evaluation techniques and tools* (pp. 261–276). London: Palgrave Macmillan.

Roy, J. (2000). The rebound effect: Some empirical evidence from India. *Energy Policy, 28*(6–7), 433–438.

Sarkar, H., & Subbarao, S. V. (1981). a short term macro forecasting model for India—Structure and uses. *Indian Economic Review, 16*, 55–80.

Saluja, M. R., & Yadav, B. (2006). *Social accounting matrix for India 2003–04*. Haryana: India Development Foundation.

Smil, V. (1998). Future of oil: Trends and surprises. *OPEC Review, 22*(4), 253–276.

Stern, M. O. (1977). A policy-impact model for the supply of depletable resources with applications to crude oil. *Energy, 2*(3), 257–272.

Shukla, P. R., Dhar, S., & Mahapatra, D. (2008). Low-carbon society scenarios for India. *Climate Policy, 8*(sup1), S156–S176.

Sugarthi, L., & Jagadeesan, T. R. (1992). A modified model for prediction of India's future energy requirement. *International Journal of Energy and Environment, 3*(4), 371–386.

Sugarthi, L., & Samuel, A. A. (2012). Energy models for demand forecasting—A review. *Renewable and Sustainable Energy Reviews, 16*(2), 1223–1240.

Sugarthi, L., & Williams, A. (2000). Renewable energy in India—a modelling study for 2020–2021. *Energy Policy, 28*(15), 1095–1109.

Shukla, P. R. (2006). India's GHG emission scenarios: Aligning development and stabilization paths. *Current Science (Bangalore), 90*(3), 384.

Tarp, F., Roland-Holst, D., & Rand, J. (2002). Trade and income growth in Vietnam: Estimates from a new social accounting matrix. *Economic Systems Research, 14*(2), 157–184.

Taylor, L. (1990). *Structuralist CGE models. Socially relevant policy analysis. Structuralist computable general equilibrium models for the developing world*. Cambridge, MA: MIT Press.

Taylor, L., & Taylor, L. (2009). *Reconstructing macroeconomics: Structuralist proposals and critiques of the mainstream*. Cambridge, MA: Harvard University Press.

TERI. (2006). National Energy Map for India: Technology vision 2030. The Energy and Resources Institute, Office of the Principal Scientific Advisor, Government of India.

The Energy and Resources Institute. (2008). *National Energy Map for India: Technology Vision 2030.*, ISBN 81-7993-064-5. New Delhi: TERI Press.

Thorbecke, E. (1992). Adjustment and equity in Indonesia. In *Development Centre of the Organisation for Economic Co-operation and Development*. Paris: OECD Publications and Information Centre [Distributor].

Vashishtha, S., & Ramachandran, M. (2006). Multicriteria evaluation of demand side management (DSM) implementation strategies in the Indian power sector. *Energy, 31*(12), 2210–2225.

Wang, T., & Watson, J. (2010). Scenario analysis of China's emissions pathways in the 21st century for low carbon transition. *Energy Policy, 38*(7), 3537–3546.

Weinberg, A. M. (1979). *Limits to energy modeling (No. ORAU/IEA-79-16 (0))*. Oak Ridge: Oak Ridge Associated Universities, Inc., TN (USA). Inst. for Energy Analysis.

York, R., Rosa, E. A., & Dietz, T. (2003). STIRPAT, IPAT and ImPACT: Analytic tools for unpacking the driving forces of environmental impacts. *Ecological Economics, 46*(3), 351–365.

Chapter 3
Evaluation of Emission Indicators Using Decomposition Analysis

Abstract The first component of the modeling framework, i.e., decomposition analysis of emission indicators, is discussed in this chapter. Emission intensity of GDP as an indicator encompasses the key energy and economic variables that are evaluated further in the other two components of this study. In this chapter, the method used for decomposition, advantages and limitations of the method, and results for a few key scenarios constructed to illustrate the method are discussed. In addition to this, we also evaluate the implications of such mitigation targets in developing countries, using India as a case study. This is done to illustrate the importance of using multiple baselines especially for developing countries while developing scenarios for energy and emission projections.

Keywords Index decomposition · Baseline emissions · Emission intensity · Nationally Determined Contributions

The first component of the modeling framework, i.e., decomposition analysis of emission indicators, is discussed in this chapter. Emission intensity of GDP as an indicator encompasses the key energy and economic variables that are evaluated further in the other two components of this study. In this chapter, the method used for decomposition, advantages and limitations of the method, and results for a few key scenarios constructed to illustrate the method are discussed.

3.1 Introduction

The greenhouse gas emissions in any country depend on a few key variables—the composition of fuels in primary energy production; the structural composition of the economy, i.e., the relative contribution of primary, secondary, and tertiary sectors to the economy; the energy intensity of different sectors; and the emission intensity of energy. An analysis of past trends reveals the contribution of each of these components to the total change in emissions and can indicate their potential role in the

© The Author(s), under exclusive licence to Springer Nature Switzerland AG 2020
T. Kanitkar, *An Integrated Framework for Energy-Economy-Emissions Modeling*,
SpringerBriefs in Environmental Science, https://doi.org/10.1007/978-3-030-18263-2_3

future. This analysis is more important for developing economies, since with unsaturated demands in all sectors the structure of the economy can vary substantially. Figure 3.1 shows the evolution of the relative contributions of three sectors—agriculture, industry, and services—to the total value added in the economy for three countries—one developed (UK) and two developing (India and China) for the period 1971–2008.

The share of the agricultural sector to total GDP is measured on the 45° line while the shares of the industrial and service sectors are measured on the *x*- and *y*-axes, respectively. For India the trend between 1971 and 2008 indicates that the reduction in the contribution of agriculture in the overall economy has been almost entirely compensated by an increase in the share of the service sector in this period. For China on the contrary this trend is markedly different, indicating a similar decline in the contribution of the agricultural sector compensated substantially by industrial sector growth. For developing economies such as India and China in the future, the decline in the share of the agricultural sector may be compensated by changes in the relative share of the industrial and service sectors, the extreme cases corresponding to increase mostly in the industrial sector or mostly in the service sector—shown by points I_I and I_S for India and points C_I and C_S for China in Fig. 3.1 for an end point taken to be 2030. The resultant emissions in 2030 from these two countries will depend significantly on the trajectory and end point of this shift in the structural

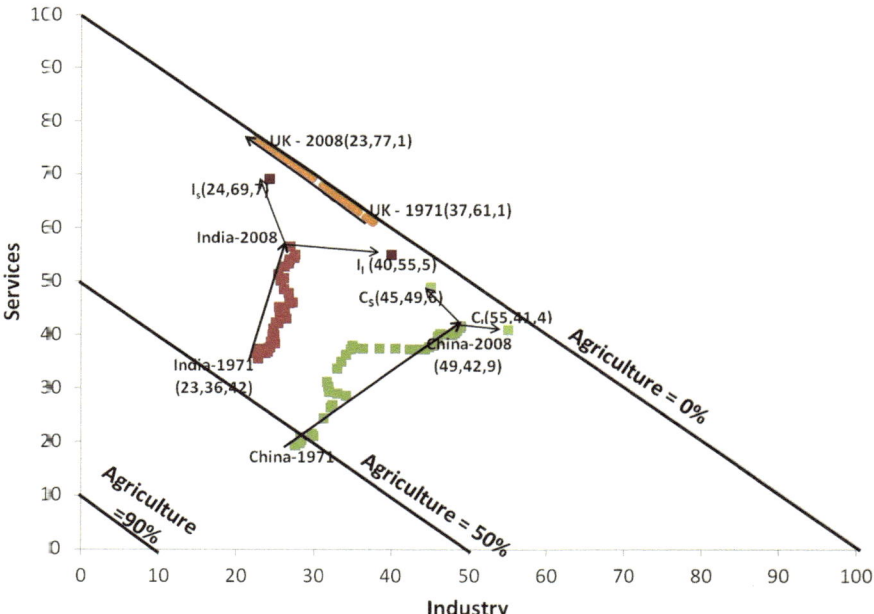

Fig. 3.1 Relative contribution of the agricultural, industrial, and service sectors to total value added in three countries—the UK, China, and India between 1971 and 2008. Data source: World Development Indicators, World Bank, 2010

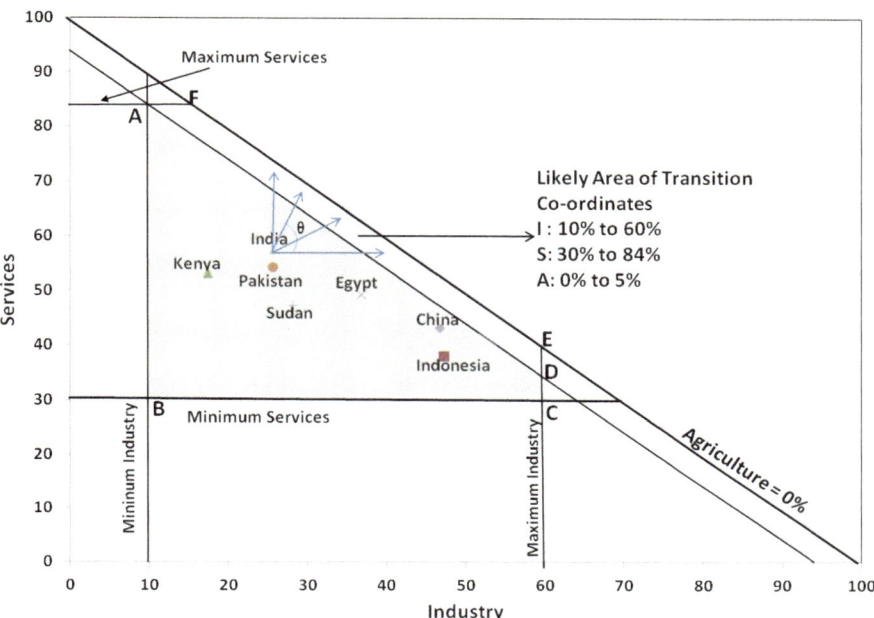

Fig. 3.2 Relative contributions to total value from the agricultural, industrial, and service sectors in seven developing economies—India, Kenya, Pakistan, Egypt, Sudan, China, and Indonesia in 2010. Data source: World Development Indicators, World Bank, 2010

composition. A similar argument can be made for several other developing countries as well, as shown in Fig. 3.2.

Shaded area A-B-C-D roughly denotes the area where most developing countries are currently situated in terms of their relative sectoral contributions. Shaded area D-E-F-A roughly denotes the area of likely transition for developing countries by 2030. The angle θ shown for India denotes the range of possible transitions. The range of available structural compositions for the future—denoted by the angle θ in Fig. 3.2—implies that a simple extrapolation of historical trajectories for economic growth may be an inaccurate measure of the potential changes in the future. We have examined this hypothesis by backcasting sectoral growth data for a range of countries. We have taken the trends for industrial growth between 1971 and 1990 to project the possible trends for the period 1991–2010. The projected values are then compared with the actual industrial growth in each country for this time period. Figure 3.3 shows the actual and projected trajectories of growth for the industrial sectors in four developing countries for the period between 1971 and 2010.

The value added is shown in the y-axis in 2000 Constant Billion USD. Given the same improvements in energy efficiency and the same changes in the composition of primary energy sources that the countries actually experienced, the projected numbers would have resulted in 23% lower emissions in China and 13% higher emissions in Brazil as compared to the actual emissions in 2010. Figure 3.3 highlights the

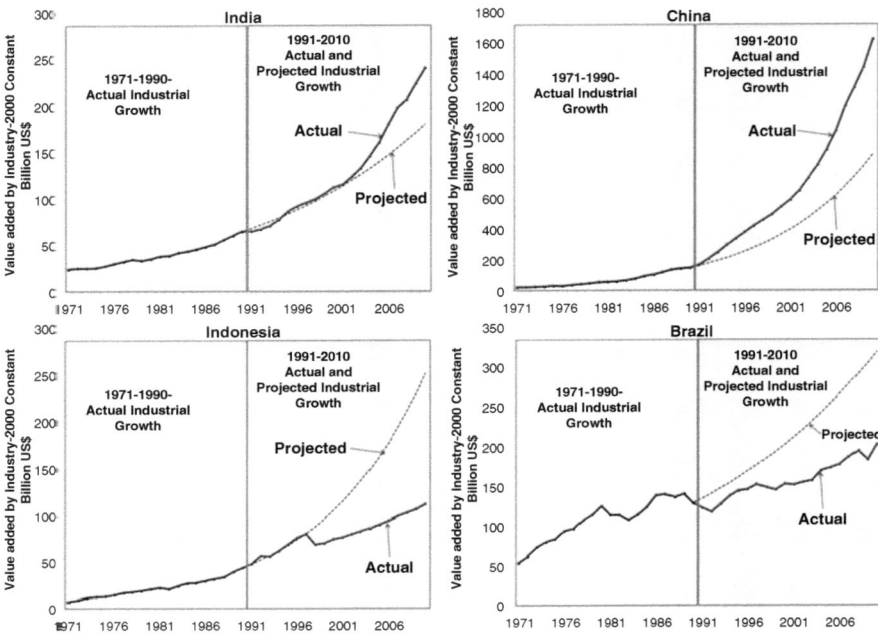

Fig. 3.3 Actual value added by the industrial sector versus projections based on historical growth rates between 1971 and 1990 for four countries—China, India, Brazil, and Indonesia. Data source: World Bank, World Development Indicators, 2010

fact that an extrapolation of the past is not a robust method to project for the future or decide a baseline for emissions. Developing economies can choose among different possible options for development depending upon their unique national priorities.

3.2   Decomposition Analysis

For a developing economy, the increase in emissions due to changes in structural composition may be offset partially by efficiency improvements and technological innovation in the energy sector. Other than the structural composition of the economy, these parameters can also affect the total emissions significantly. Using the decomposition method, the total emissions in any country can be expressed as a function of all these parameters as shown in Eq. (3.1):

$$E = \sum_i I_i \times C_i \times S_i \times G \tag{3.1}$$

Decomposition analysis is used to evaluate the historical contributions of three main factors to the economy—(i) the energy intensity of GDP—energy consumed per unit GDP in each sector; (ii) the emission intensity of energy—carbon dioxide emissions per unit of energy; and (iii) the structure of the economy—relative contributions from each of the three sectors. The multiplicative and additive log mean Divisia index (LMDI) techniques for decomposition were used as they have been shown to be the most useful for policy purposes due to their theoretical foundation, adaptability, and ease of use, Ang (2005). The calculated LMDI values are based on Eqs. (3.1–3.4). For multiplicative LMDI, the effect of the kth component can be calculated using

$$\text{LMDI}k = \exp\left\{\sum \frac{L\left(V_i^T, V_i^0\right)}{\sum L\left(V_i^T, V_i^0\right)} \times \ln\left(\frac{k_i^T}{k_i^0}\right)\right\} \tag{3.2}$$

where k is the component for which the index is calculated, i.e., the change in emission intensity due to structural changes or energy intensity improvements in the economy between year 0 and year T, and V_i is the energy or emission indicator under evaluation—in this case the emission intensity of GDPand

$$L(a, b) = \frac{a - b}{\ln(ab)} \tag{3.3}$$

For additive LMDI,

$$\text{LMDI}k = \sum \left(L\left(V_i^T, V_i^0\right) \times \ln\left(\frac{k_i^T}{k_i^0}\right)\right) \tag{3.4}$$

The arithmetic mean Divisia index (AMDI) has also been calculated for comparison using the following equations:

$$\text{AMDI}k = \exp\left\{\sum W_i \times \ln\left(\frac{k_i^T}{k_i^0}\right)\right\} \tag{3.5}$$

where

$$W_i = \frac{\left(\frac{V_i^T}{V^T}\right) + \left(\frac{V_i^0}{V^0}\right)}{2} \tag{3.6}$$

3.2.1 Decomposition Analysis for India

For the purpose of the analysis, the Indian economy is disaggregated into three sectors, viz., industry, services, and agriculture. The value added in industry includes value added in mining, manufacturing, construction, electricity, water, and gas. The value added in services includes value added in wholesale and retail trade (including hotels and restaurants), transport, and government, financial, professional, and personal services such as education, health care, and real estate services. The value added in agriculture includes forestry, hunting, and fishing, as well as cultivation of crops and livestock production. The decomposition indices are evaluated for the time period between 1971 and 2008.

The arithmetic mean Divisia index (AMDI) method (Ang 2004) has been used for comparison. The indices are evaluated for four different time periods based on a shift in the trends in emission intensity shown in Fig. 3.4.

Figure 3.4 shows the trends in emission intensity of GDP, emission intensity of energy, and energy intensity of GDP. The emission intensity of GDP shows distinctive trend breaks which are used to divide the entire time period into four sub-periods, viz., 1971–1978, 1978–1987, 1987–1995, and 1995–2008. The AMDI and LMDI indices for these four time periods are shown in Table 3.1.

The annual change in each component is shown in Table 3.2.

The emission intensity of GDP shows a marginal reduction between 1971 and 1978 (0.1% per year). Almost the entire reduction in emission intensity achieved in this time period can be attributed to the reduction of energy intensity of GDP, i.e., to an improvement in energy efficiency. The evaluation of the LMDI and AMDI indices shows that in the subsequent time periods also the reduction in energy intensities has played a significant role in reducing emission intensities. Table 3.3

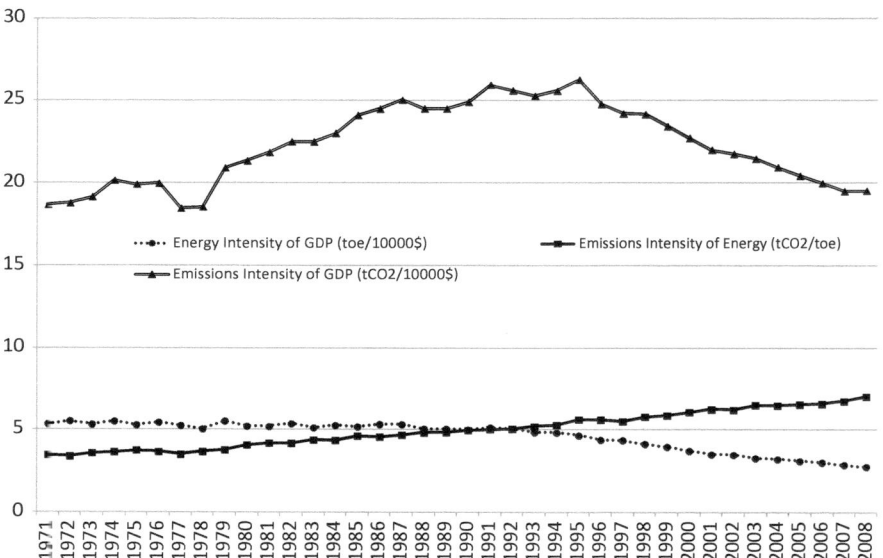

Fig. 3.4 Shifts in emission intensity trends between 1971 and 2008. Data source: World Bank Country Indicators

Table 3.1 LMDI and AMDI for four time periods

	1971–1978		1978–1987		1987–1995		1995–2008	
	LMDI	AMDI	LMDI	AMDI	LMDI	AMDI	LMDI	AMDI
Multiplicative								
Total effect	0.99	0.99	1.349	1.349	1.050	1.050	0.744	0.744
Structural effect	1.06	1.06	1.068	1.068	1.056	1.056	1.009	1.009
Energy intensity effect	0.89	0.89	0.988	0.989	0.784	0.784	0.612	0.612
Emission intensity effect	1.05	1.05	1.279	1.276	1.268	1.267	1.205	1.205
Additive								
Total effect	−0.00001	−0.00001	0.00065	0.00065	0.00012	0.00012	−0.00067	−0.00067
Structural effect	0.00197	0.00010	0.00231	0.00014	0.00271	0.00014	0.00231	0.00002
Energy intensity effect	0.00167	−0.00021	0.00216	−0.00002	0.00204	−0.00062	0.00142	−0.00113
Emission intensity effect	0.00196	0.00009	0.00283	0.00054	0.00329	0.00061	0.00275	0.00043

Table 3.2 Annual change in emission and energy indicators

	1971–1978	1978–1987	1987–1995	1995–2008
	Change per year (%)	Change per year (%)	Change per year (%)	Change per year (%)
Total emission intensity	−0.10%	3.88%	0.62%	−1.97%
Change in structure	0.83%	0.75%	0.70%	0.07%
Change in energy intensity of GDP	−1.54%	−0.13%	−2.70%	−2.99%
Change in emission intensity of energy	0.74%	3.10%	3.35%	1.57%

Table 3.3 Changes in the energy intensity of GDP for industry, services, and agriculture for four time periods

	1971–1978	1978–1987	1987–1995	1995–2008
Industry	−1%	0.1%	−3.3%	−2.7%
Services	−3%	−1.7%	−1.4%	−4.6%
Agriculture	6%	10.6%	9.5%	0.1%

shows the trends of energy intensity of GDP for the three sectors under consideration for all four time periods.

The highest reduction in the energy intensity of GDP has been registered in the service sector, followed by the industrial sector in the last two time periods (1987–2008). The improvements in energy efficiency in the industrial and service sectors have been responsible for the reduction in the overall emission intensity of GDP of the economy.

3.2.2 Drivers of Energy and Emission Intensities

The energy intensity per unit of value added in the agricultural sector has increased between 1970 and 2008. There has been a steady growth in groundwater irrigation in this period. Between 1993 and 2001, the net area irrigated through groundwater increased by about 25% as opposed to a 24% decline in the net area irrigated through surface water. Apart from the increase in the use of pump sets, the sector also has experienced a steady increase in the level of the mechanization for other operations as well (e.g., the use of tractors as opposed to livestock for ploughing operations). However, in spite of these factors, the increase in the energy intensity of GDP in agriculture between 1971 and 2008 is not constant, as expected. There is a sharp reduction in energy intensity around 1998–1999 and a relative stable trend thereafter. This may indicate a sharp improvement in efficiency of pump sets and other farm equipment in this period. However, a more likely explanation of this trend can be found in the change in the manner of accounting of power supplied by the distribution utilities. The Regulatory Commissions Act was passed in 1998 and in most states the

State Electricity Regulatory commissions (SERCs) were set up, after which most state power distribution utilities were required to meter agricultural transformers. This led to a shift in the accounting of agricultural consumption, as a large amount of electricity consumption previously considered as agricultural was in fact attributable to technical and commercial losses of the utility (Mahalingam 2002).

It is important to note that the analysis of energy and emission intensities does not include noncommercial primary energy supplied for residential consumption. Commercial primary energy as well as secondary supply for residential consumption is included in industrial consumption as it considers electricity, gas, and other processing of petroleum products. However, noncommercial energy continues to play a significant role in India's energy sector (24% of total energy supply in 2010) even though it has reduced from a relatively higher share in 1971 (61%). A large percentage of this energy supply (79%) in 2010 was consumed by the residential sector. As the economy develops further, it is reasonable to assume that the percentage of noncommercial energy (mainly consisting of biomass and other biofuels—considered carbon neutral) will reduce and the resultant increase in the demand for commercial energy will be reflected in the energy consumption within the "industrial" sector.

The share of fossil fuel energy in the total primary energy mix has increased steadily from about 37% in 1971 to 71% in 2008. Combustible renewable and waste[1] have decreased as a percentage of total primary energy to constitute about 26% of total energy use in 2008. A large part of this category of "combustible renewable and waste" comes from traditional fuels such as solid biomass and biogas. A reduction in the use of these traditional fuels as a percentage of total fuel use signals a shift from traditional fuel sources to more modern sources of fuel. This shift towards modern fossil fuels is likely to continue if the experience of other countries is to be taken as indicative. Fossil fuels play a much larger role in total energy supply in other developed as well as developing economies as compared to India as shown in Fig. 3.5.

The share of alternative and nuclear energy as a percentage of total energy supply has increased only in developed economies such as the United States and Germany. Alternative and nuclear energy in India has played a very small role over the past three decades increasing only marginally from 1.5% to 2.5% as a percentage of total primary energy supply. In the power sector alone, the share of fossil fuels increased from 60% to 70% (in terms of installed capacity) and 60% to 80% in terms of total power generation between 1977 and 2008.

The share of hydropower has reduced substantially from 38% in 1977 to 21% in 2008 in terms of installed capacity and even more substantially from 36% in 1977 to 15% in 2008 in terms of the total energy generated. The average rate of annual capacity addition from various power generation technologies between 1980 and 2008 is shown in Table 3.4.

---

[1]Combustible renewables and waste comprise solid biomass, liquid biomass, biogas, industrial waste, and municipal waste.

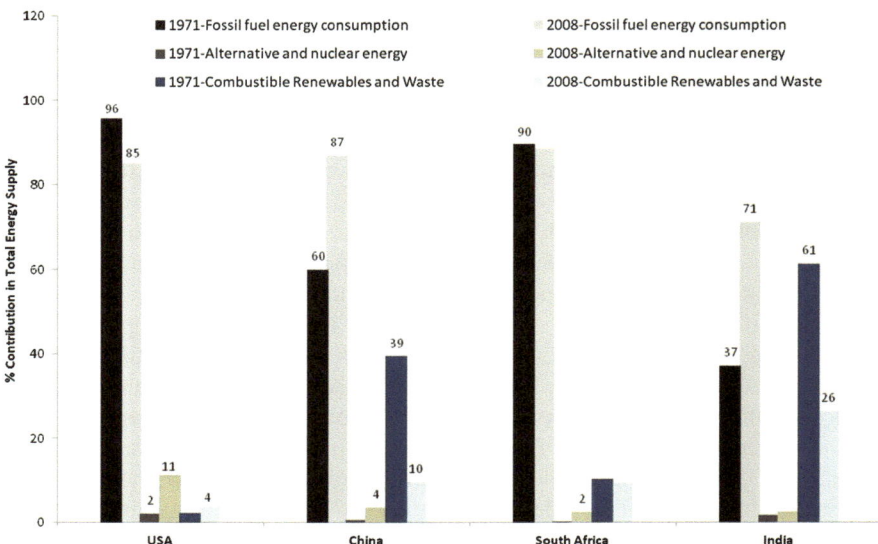

Fig. 3.5 Contribution of fuel sources in total energy supply—comparison of four countries. Data source: IEA, World Energy Outlook, 2010

Table 3.4 Average annual rate of capacity addition in the power sector between 1980 and 2008

	Coal	Nuclear	Natural gas	Hydro	Diesel	Renewable	Total
Total capacity added between 1980 and 2008 (GW)	70	3	19	25	9	11	137
Average annual rate of capacity addition	6%	7%	17%	4%	9%	22%	6%

The average rate of capacity addition is the lowest for hydropower. Comparatively longer periods of construction, a change in the structure of finance for the sector, as well as increasing concerns on provision of adequate resettlement for the populations displaced due to hydro projects have a significant role to play in the low rates of capacity addition in the sector in the period post-1980 (Oud 2002). However, the total hydro potential in the country as estimated in the Integrated Energy Policy is about 150 GW (GoI 2006a, 2006b) and estimates for the future show a revival of interest in hydropower with increasing costs of energy from fossil fuel sources as well as concerns of climate change mitigation (Bartle 2002). New renewable energy technologies have seen the highest rates of growth in this time period. The share of new renewable sources such as wind and solar energy has increased in the power sector, more substantially in the past 5 years. However these constitute about 7% of the current installed capacity in the country and only about 3% of the total energy generation.

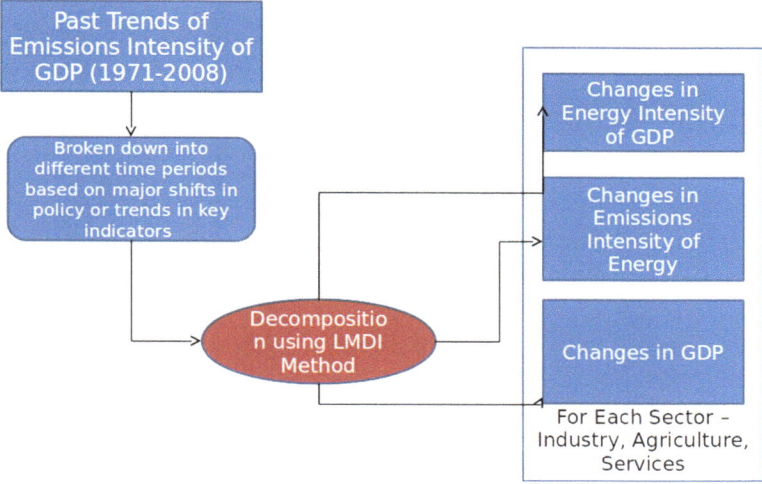

Fig. 3.6 Illustration of decomposition analysis and construction of scenarios for the future

The contribution of coal in total power generation shows a marginal decline between 2000 and 2008. However, part of this decreased contribution is compensated by an increase in the share of natural gas. The reduction in the share of hydropower in overall power generation, the shift from the use of traditional fuels to more carbon-intensive fossil fuels, as well as the lack of a substantial change in the scenario of either new and renewable energy or nuclear power in the energy sector have led to a positive contribution to the overall emission intensity of energy.

The insights from the decomposition analysis are used to construct economic scenarios and energy scenarios for the future. The methodology applied in using the analysis of past trends for scenarios for the future is illustrated in Fig. 3.6.

3.3 Scenarios for Target Year 2030

In this analysis, we estimate future emissions including scenarios for structural change in the economy. The potential change in emissions in the future and the resultant changes in mitigation efforts required in India are evaluated for two basic scenarios. The main variable under exploration in this analysis is the structural composition of the economy. The potential trajectory of future economic choices for developing economies can be measured by the angle θ (refer Fig. 3.2), where

$$\tan \theta = \frac{\Delta S}{\Delta I} \tag{3.7}$$

where ΔS denotes the change in the contribution by the service sector to the total economy and, and ΔI denotes the change in the contribution by the industrial sector to the total economy. $\theta \geq 90°$ implies that the entire change in the economic structure is a result of the growth in the industrial sector replacing agriculture as the economy develops whereas $\theta \leq 0°$ implies that the entire change in the economic structure is a result of the growth in the service sector.

3.3.1 Construction of Economic Baselines

Two scenarios for India capture two possibilities in the economic choices for the future for India, viz., (i) high growth in the service sector and (ii) high industrial growth.

1. Scenario 1—high growth of the service sector—The service sector in India is currently growing at a faster rate compared to the other sectors and the decline of the contribution of agriculture in the economy has mostly been compensated by the increase in the share of the service sector. Scenario 1 extrapolates a similar trend to the future—the growth rate of each sector in the past 5 years continues into the future (2.8% p.a. for agriculture, 6.6% p.a. for industry, and 8.1% p.a. for services). In this scenario the service sector accounts for about 69% of the total value added in the economy by 2030 and the overall rate of economic growth is 7.1% p.a.
2. Scenario 2—high growth of the industrial sector—There is a large infrastructure deficit in developing countries such as India. Government plans project an increase in the growth of heavy industry and primary energy sectors to overcome this deficit—targeted growth rates of 9.6–10.9% in the industrial sector and 8.5–9% in the energy sector (GoI 2011a, 2011b). Scenario 2 presents a scenario of high industrial growth where the total value added by the industrial sector by 2030 accounts for almost 40% of the total value added in the economy. In this scenario the industrial GDP is estimated to grow at the rate of 9.1% p.a. and the overall economic growth rate is 6.7% p.a.

The scenarios are further developed by introducing potential changes in technology (improvements in efficiency) and the introduction of green technologies in the energy sector. This is captured by three possibilities—(i) the sectoral values for energy intensity and emission intensity remain constant. This would imply that no further improvement in energy efficiency in any of the sectors is possible. For agriculture it will also imply that the level of mechanization will undergo no significant change. This scenario of "frozen efficiency" can be used to provide a benchmark or baseline against which other scenarios may be evaluated. (ii) The trends in energy intensity of GDP and emission intensity of energy seen in the last time period continue into the future. This would imply that for agriculture there would be a gradual increase in the level of mechanization and for the industrial and service sector efficiency improvements would take place at the same rate without saturation between 2009 and 2030. However, constant trends for emission intensity of energy would imply that there would also be a steady increase in the share of carbon energy in the fuel mix for this time period counterbalancing the achievements in energy efficiency. In this analysis we present results for scenarios in which the energy intensity of the gross domestic product (GDP) and the emission intensity of GDP follow business as usual trajectories. The energy intensity of GDP reduces by 2.8% p.a. till 2030. The emission intensity of energy increases by 1.7% p.a. even with a steady increase in the use of renewable energy sources. The resulting emission trajectories from these scenarios provide a range of baselines over which mitigation requirements would then have to be calculated.

3.3.2 Mitigation Targets for Developing Countries

As part of the Nationally Appropriate Mitigation Actions (NAMAs) in an earlier phase and as Intended Nationally Determined Contributions (INDCs) as part of the Paris Agreement, many developing countries have proposed voluntary mitigation actions (UNFCCC 2011; UNFCCC 2015). The actions proposed by developing countries may be classified as follows:

1. Reduction from business as usual emissions: Many developing countries have proposed targets that specify reductions in carbon emissions from a business-as-usual baseline. For example, South Africa has proposed to implement mitigation actions that would reduce its emissions by 34% below the "business-as-usual" emissions by 2020 and by 42% below business-as-usual emissions by 2025. Brazil, Mexico, and South Korea are some of the other emerging economies that have used this framework for setting mitigation targets.
2. Reductions in the emission intensity of GDP: China and India have proposed mitigation targets in terms of a reduction of the emission intensity per unit of GDP. India has proposed a 20–25% reduction in its emission intensity of GDP from 2005 levels by 2020 and 33–35% reductions with respect to the same baseline by 2030.
3. Reduction in absolute flows of emissions with respect to a base year: Although this framework of setting mitigation targets is usually considered to be applicable only to developed countries, some developing countries have also proposed an absolute reduction in emissions. For example Indonesia has proposed a 26% reduction in emissions by 2020, expected to be achieved mainly through management of land use, and Antigua and Barbados has proposed a 25% reduction below 1990 levels by 2020. It is not possible for all developing countries to adopt such targets at this stage, given the requirements of poverty alleviation and development in most of these countries. However, as some developing countries have proposed mitigation targets of this kind, we do include them in our analysis.
4. Sectoral mitigation actions—Many developing countries have proposed specific actions to be taken within each sector instead of proposing an overall emission reduction target, e.g., Argentina, Armenia, and Columbia. As most of these

targets are not quantified in terms of emissions it is difficult to compare this framework with the other methodologies discussed here. Therefore this class of mitigation targets is excluded from the analysis in the work.

The first three types of targets are applied to the baselines constructed for India to calculate the amount of effort required to achieve emission reduction. Three national mitigation targets for India are considered—(i) the emission intensity of GDP to be reduced by 25% as compared to current (2008) levels, (ii) the annual emission flows to be reduced by 25% by 2030 relative to a business as usual baseline, and (iii) the annual emission flows to be kept constant between 2008 and 2030.

3.4 Results and Discussion

The difference in mitigation effort for the two scenarios with different targets is measured by the differences in the corresponding cumulative emissions over the reference period, from 2009 to 2030.

The mitigation efforts implied by the areas marked in Fig. 3.7 are shown in Table 3.5 to provide the magnitude of difference made by choosing two alternative routes for development. The baseline cumulative emissions between 2009 and 2030 are estimated to be about 70 $GtCO_2$ for Scenario 1 and 88 $GtCO_2$ for Scenario 2.

In the event of a higher contribution in the future by the industrial sector, the mitigation effort required would be higher by about 19–38% for a range of mitigation targets. For India, the magnitude of effort required to achieve its target of 25% reduction in the emission intensity of GDP, from 2005 levels, by 2020 will be different for the two scenarios. In Scenario 1, it is possible to achieve the target by a substantial improvement (~3.7% p.a.) in energy efficiency (as compared to the currently projected rates of efficiency improvement of 2–2.5%) without any significant change in the mix of primary energy sources. For Scenario 2, the rate of efficiency improvement required would be 5.1%. This is based on an assumption of an increase in the emission intensity of energy (following a business-as-usual trajectory of increased share of fossil fuels in the fuel mix). The rate of energy efficiency improvements can be reduced if the share of renewable energy sources in the fuel mix increased but either option has substantial costs attached to it. A similar analysis undertaken for the analysis of the Intended Nationally Determined Contributions (INDC) submitted by the Government of India shows that the target of 33–35% reduction in emission intensity will require additional effort if the structure of the economy were to change with industry playing a larger role.

While a consistent effort towards developing a framework for a global climate agreement is important, developing economies will have to assess their unique national circumstances in order to allow for a range of development options for the future. It is necessary to devise an alternative architecture within which both the goals of mitigation and the freedom of nation states to choose their economic futures are protected.

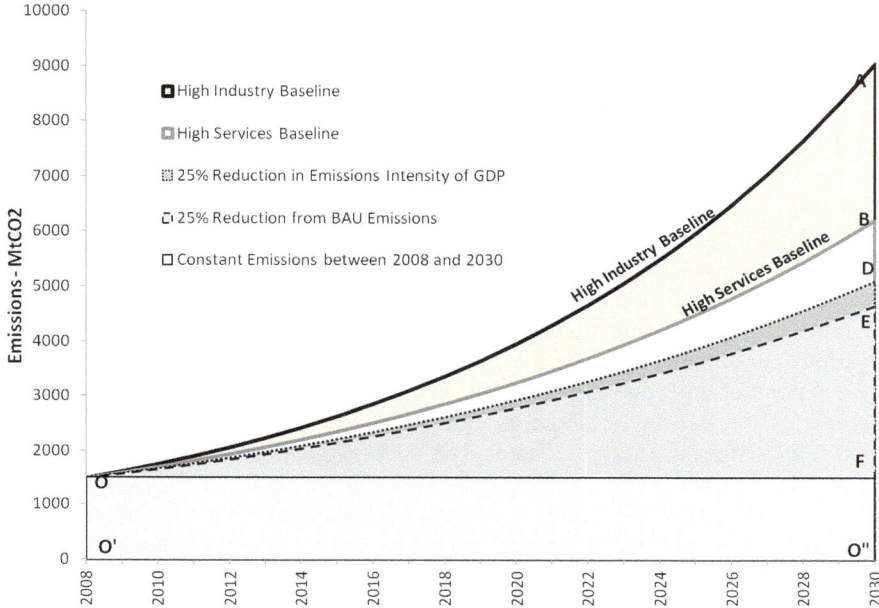

Fig. 3.7 Mitigation effort required based on two baselines—(i) high industrial growth—curve OA and (ii) high service growth—curve OB. Curve OD shows the emissions for a mitigation scenario of 25% reduction in emission intensity of GDP by 2030. Curve OE shows the emissions for a mitigation scenario of 30% reductions by 2030 from a business-as-usual baseline. Curve OF shows the emissions for a mitigation scenario wherein the emissions are not allowed to increase—constant emissions between 2008 and 2030

Table 3.5 Mitigation effort required for Scenarios 1 and 2

	Cumulative emissions between 2009 and 2030 (GtCO$_2$)	Emission reduction required—measured as difference between baseline trajectories and mitigation trajectories (GtCO$_2$)	
		From high service baseline (70 GtCO$_2$)	From high industry baseline (88 GtCO$_2$)
Mitigation target 1–25% reduction in the emission intensity of GDP by 2030	67	3 GtCO$_2$ (4% reduction from baseline)	21 GtCO$_2$ (24% reduction from baseline)
Mitigation target 2–25% reduction from business-as-usual emissions by 2030	59	11 (16% reduction from baseline)	29 (33% reduction from baseline)
No increase in emissions between 2009 and 2030[a]	33	37 (53% reduction from baseline)	54 (61% reduction from baseline)

[a]An absolute reduction in emissions is not a feasible target for India. However the calculation for such a case is presented here to provide a benchmark for minimum emissions

3.5   Conclusion

The analysis presented here indicates that the structural composition of the economy matters should be explicitly considered when setting mitigation targets for developing countries or indeed even while constructing scenarios for the future. The level of mitigation effort required is higher in case the future economic growth profile involves a higher share of industry. Mitigation targets measured from business-as-usual baselines, without the mention of a potential economic structure, nevertheless contain an implicit assumption of future economic growth trajectories. A more robust framework would consider multiple baselines for any country and as a result a range of potential mitigation targets. By providing information about the role that various contributing factors play in the reduction or increase in emissions at any given point of time, decomposition analysis provides a robust basis for constructing scenarios for the future. From analysis of this nature, the range of values for indicators such as energy efficiency or sectoral growth for each sector can be determined. Many models tend to use continuously improving energy efficiencies or business-as-usual economic growths to construct scenarios for the future. A more thorough analysis of past trends provides a measure of ground truthing and robustness to the values used for some of these parameters for the future. It provides a robust basis to decide upper and lower limits for values that variables can assume in the future. An example of this method used for such a purpose in conjunction with other models is discussed in Chap. 6.

References

Ang, B. W. (2004). Decomposition analysis for policymaking in energy: which is the preferred method? *Energy Policy, 32*(9), 1131–1139.

Ang, B. W. (2005). The LMDI approach to decomposition analysis: a practical guide. *Energy Policy, 33*(7), 867–871.

Bartle, A. (2002). Hydropower potential and development activities. *Energy Policy, 30*(14), 1231–1239.

GoI. (2006a). Government of India, Office of the Principal Scientific Adviser and The Energy and Resource Institute. In *National Energy Map for India: Technology Vision 2030*. New Delhi: TERI Press.

GoI. (2006b) Planning Commission, Government of India. Integrated Energy Policy: Report of the Expert Committee. New Delhi.

GoI. (2011a) Planning Commissions, Government of India. 12th Five Year Plan. Retrieved December 9, 2015, from http://12thplan.gov.in/index.php

GoI. (2011b) Planning Commissions, Government of India. Interim Report of the Expert Group on Low Carbon Strategies and Inclusive Growth. Retrieved December 6 2015, from http://planningcommission.nic.in/reports/genrep/Inter_Exp.pdf

Mahalingam, S. (2002). The great power robbery. *Frontline, 19*(06), 2002.

Oud, E. (2002). The evolving context for hydropower development. *Energy Policy, 30*(14), 1215–1223.

United Nations Framework Convention on Climate Change (2011) Compilation of Information on Nationally Appropriate Mitigation Actions to be Implemented By Parties not Included In Annex I to the Convention, FCCC/AWGLCA/2011/INF.1. Retrieved December 5 2015, from http://unfccc.int/resource/docs/2011/awglca14/eng/inf01.pdf

United Nations Framework Convention on Climate Change (2015) Report of the Conference of the Parties on its twenty-first session, held in Paris from 30 November to 13 December 2015, FCCC/CP/2015/10/Add.1. Retrieved May 8 2019, from https://unfccc.int/process/conferences/pastconferences/paris-climatechange-conference-november-2015/paris-agreement.pdf

Chapter 4
Optimal Pathways for Power Supply

Abstract The second component of the integrated modeling framework is evaluating the optimal energy supply pathways for a particular region for a particular time given multiple constraints on resources, capital, emissions, etc. Various methods can be used to estimate energy demand. One such method used in this modeling framework is an input-output analysis that estimates economy-wide energy demand. This is discussed in Chap. 5. The next step after estimating demand is to evaluate ways in which this demand can be met. This includes estimating the appropriate fuel mix to satisfy energy needs. In this chapter a model is presented for the power sector that takes into consideration some of the key characteristics of the sector and fuel technologies. The objective is to arrive at estimates for the optimum fuel supply pathway for the power sector for a given demand. This chapter includes a detailed description of the power sector model built using the GAMS (General Algebraic Modeling System) platform and a few scenarios constructed for the purpose of illustrating the working of the model.

Keywords Constrained optimization · GAMS · Fuel-mix · Generation expansion planning · Carbon budget · Least-cost energy pathways

4.1 Introduction

The second component of the integrated modeling framework is evaluating the optimal energy supply pathways for a particular region for a particular time given multiple constraints on resources, capital, emissions, etc. Various methods can be used to estimate energy demand for a given time. One such method used in this modeling framework is an input-output analysis that estimates economy-wide energy demand. This is discussed in this chapter. The next step after estimating demand is to evaluate ways in which this demand can be met. This includes estimating the appropriate fuel mix to satisfy energy needs. At any given time, there are existing energy supply systems with specific technical and financial characteristics. To supply increasing energy requirements, newer energy supply

© The Author(s), under exclusive licence to Springer Nature Switzerland AG 2020   45
T. Kanitkar, *An Integrated Framework for Energy-Economy-Emissions Modeling*,
SpringerBriefs in Environmental Science, https://doi.org/10.1007/978-3-030-18263-2_4

options have to be added to the system while some existing systems will be decommissioned. Each fuel supply technology has characteristics of its own—in terms of both technical peculiarities and capital requirements. In addition to this there are constraints imposed by the system as well as the environment in which these technologies get deployed. For example, installing and commissioning a nuclear power plant takes a longer time as compared to, say, a gas-based plant. This is in part because of the difference between the two technologies themselves. However part of the reason is also because installing a nuclear energy plant would require more clearances, a longer process of making sure that the plant is acceptable to the local community, etc. There are also constraints on the way in which these technologies can function together in a system, also depending in part on the characteristics of the energy demand. For example the size of the manufacturing sector would dictate the amount of base load power required in any system.

In the work presented here, a model is built for the power sector that takes into consideration some of the key characteristics of the sector and fuel technologies. The objective is to arrive at estimates for the optimum fuel supply pathway for the power sector for a given demand. In this chapter, energy demand is considered as an independent input to the power sector model. But it is useful to recall that in the integrated framework there is a feedback loop between the power sector model and the input-output model. The energy demand estimated in the input-output model is an input to the power sector model. The capital investment required for a particular fuel mix which is an output of the power sector model is in turn fed back into the input-output model. The power sector model is built using the GAMS (General Algebraic Modeling System) platform, described in the next sections, and a few scenarios are constructed for illustration.

4.2 Data and Assumptions

The power sector is the largest component of the energy sector as a whole and in a developing economy, of considerable importance to the development of a country's infrastructure. In India, the power sector has been growing at an average rate of about 5% per annum for the past 5 years. Therefore, the power sector has been considered for analysis here in order to demonstrate the implications of different fuel supply mixes. The model discussed here can be described as a modified version of a generation expansion planning model wherein energy costs are minimized subject to constraints on both mineral and atmospheric resources. The input parameters and assumptions built into the model along with data sources used for each major input parameter are explained here.

Table 4.1 Electricity demand in 2035 using different methods

	Electricity demand in 2035 (kWh/person/year)
Using an IPAT analysis—assuming a GDP growth rate of 6.9% p.a. and an annual reduction of 2% in the energy intensity of GDP	1913
Using a SAM—for an overall GDP growth rate of 6.9%	1919–2318
Normative values based on correlating HDI and electricity use—correlation with HDI of mid-level developed countries—~ 0.8–0.85	4500

4.2.1   Energy Requirements

The values for energy requirements in a target year can be obtained through a variety of methods, viz., input-output analysis, a simple IPAT analysis, or a statistical analysis correlating energy and development across a host of countries. The electricity requirement in 2035 (in kWh/person/year) using a variety of such methods is shown in Table 4.1.

The per capita electricity use for India in 2010 was 705 kWh/person/year (CEA 2011). The per capita electricity consumption of mid-level developed countries ranges from 4500 to 6000 kWh/person/year (World Bank 2012). An assumption of the possibility of decoupling growth and human development entirely from fossil fuel use would involve speculation that cannot be justified against the goals of achieving high levels of human development at this stage. The technological basis of providing the energy requirements to achieve high levels of development may of course change in the future, but there is no reasonable way in which to predict these changes. The per capita electricity consumption of mid-level developed countries (such as Portugal, which shows a lower consumption of electricity for the same level of human development as compared to other mid-level developed countries with a similar profile) can therefore be used as a benchmark value to be achieved by a particular year. However there may be some room to reduce this number based on technology and efficiency improvements. There are various projections for energy demand also given by the Planning Commission and the Ministry of Power, Government of India, and those obtained from input-output analysis or other similar analyses, which together with the previous methodology provide a range of values that may be used to create scenarios against which the power sector model may be constructed. Thus, by choosing a benchmark for the achievable best, a target for energy consumption to be achieved in a given time period in the future can be arrived at. The electricity demand curve is then constructed by the model based on the specified target values.

When used in conjunction with the other models in the integrated modeling framework, the energy demand estimated by the input-output analysis is used as an input to this model. In the scenarios constructed here, we have taken a value of 2500 kWh/person/year to be achieved by 2030 and 3500 kWh/person/year to be achieved by 2050 to illustrate the operation of the model. The value of 2500 kWh/person/year is estimated to be the electricity demand in 2030 in the document submitted by the Government of India to the UNFCCC (United Nations Framework Convention on Climate Change) as part of India's INDC (Intended Nationally Determined Contribution), now the Nationally Determined Contribution (NDC) post-ratification in October 2016.

4.2.2   Power Supply

India is relatively diverse in terms of the fuel sources that supply the power demand. However, the share of coal-based power generation is the highest. Of the existing 71 GW of coal-based power generation capacity in India existing in 2009, about 35 GW would still be operating in 2030 and all of these plants would be decommissioned by 2047 (based on the estimates given by the Ministry of Power, Government of India). Similarly, the existing nuclear power capacity and gas-based generation in 2009 are estimated to be decommissioned by 2045. 30 GW of existing hydro-based capacity however is expected to still be available till 2050 (CEA 2009). Given the expected lifetimes for existing power plants and the total capacity that will be available from currently operational plants, the gap between the total demand between 2010 and 2050 will be met by additional generation from a variety of sources.

The distribution of electricity sources in India currently is accounted for by two factors: (i) cost of electricity generation and (ii) resource availability. India has high amounts of coal reserves. Gas has recently started to acquire greater significance in power generation; however demand for natural gas from other sectors (fertilizer production for example) is likely to limit the role that it can play in the power sector in the future. However, most studies of the power sector in India project an increase in the share of gas-based generation in the medium term (GoI 2006). A large portion of the hydro reserves are also yet to be tapped; India has a total potential of about 150 GW of which only about 41 GW has been tapped so far (33 GW by 2009). However, the capacity factors of hydroelectric plants are low compared to coal plants and a large part of the new hydro capacity is expected to be of the "run-of-the-river" variety, with lower capacity factors. Electricity generation from renewable energy sources in the country has increased considerably in the last decade and in 2010 these accounted for 10% of total installed capacity and 5% of total electricity generation. A large percentage of this increase can be attributed to an increase in wind energy generation. The total energy generation potential and average capacity factors (actual values for 2009 and expected values in 2030 and 2050) for each fuel source are shown in Table 4.2.

The sources of electricity that will meet the energy demand are constrained by many factors such as resource constraints (e.g., availability of coal, gas, uranium), as well as constraints on factors such as water and land. The second category of constraints is not considered in this model. The resource constraints estimated by

Table 4.2 Fuel-wise capacity factors and total energy potential[a]

	Actual average capacity factor/plant load factors (2009)	Expected average capacity factor/plant load factors (2030)	Total energy potential[b]
	%	%	GW
Coal	75%	85%	600
Nuclear	70%	80%	65
Hydro	38%	50%	150
Gas—CCGT	70%	80%	100
Diesel	20%	20%	15
Wind	19%	18%	300
Small hydropower	25%	30%	15
Biomass	65%	70%	70
Solar photovoltaic	19%	20%	NA
Concentrated solar power	23%	23%	NA
Municipal solid waste	60%	60%	25

[a]Sources: Ministry of New and Renewable Resources, Report of the Lawrence Berkley National Laboratory, Ministry of Power
[b]The total energy potential refers to the maximum energy flux that can technically be generated using each technology

various agencies can vary quite widely and have a significant impact on the outcomes of the kind of fuel mixes that may be possible in the future. In the model, this input can also be modified as better estimates become available.

The actual amount of generation and installed capacity need to be higher than consumption to account for losses in the transmission and distribution network (T&D losses), as well as auxiliary consumption by the plants. The T&D losses in India are high (25% of total electricity generated is lost in transmission and distribution) partly because India has the highest length of low-tension distribution lines with very low density of domestic consumers at the tail end especially in rural areas as well as low usage of electricity by domestic consumers. The model allows the user to input a reduction of T&D losses from 25% in 2010 to 20, 15, or 10%, by 2030. Apart from the T&D losses and auxiliary consumption, a spinning reserve of 10% is used as standby in case of unforeseen failures in power plants.

4.2.3 Cost of Electricity

The aim of the model is to provide a potential "least-cost" fuel mix while simultaneously ensuring that the future emission constraint for India will not be violated. Therefore, the cost of electricity generation and potential trajectories of this cost for each fuel source are important inputs. The cost of generation for each technology includes three major components—(i) capital costs, (ii) operation and maintenance (O&M) costs, and (iii) fuel costs. It is possible to determine the current levelized cost of generation using these components for each technology. However, to project the behavior of energy cost in the possible future involves many assumptions (Yeh and Rubin, 2012). Capital costs may vary based on actual increase in material costs, changes in market conditions, or developments in the regulatory and policy framework within which these costs are determined. Similarly, O&M costs may vary based on changes in labor costs in the particular country of analysis and the fuel costs may vary based on either actual changes in the cost of fuel extraction or variations in market conditions. Using cost projections to reliably determine financial burdens therefore is a challenging task. Despite these uncertainties however, it is possible to determine the levels of additional financial burdens that will have to be incurred given increasing constraints in allowed emissions, in a robust manner, if the cost assumptions are maintained constant for the base as well as subsequent scenarios. It is also possible to verify the robustness of the results by generating a series of scenarios for varying cost assumptions. In our model we vary the cost inputs, thereby generating different scenarios for a range of carbon constraints and energy demands. In the results presented in this analysis, the current capital and energy costs for each technology are estimated based on norms established by the Central Electricity Regulatory Commission, India (CERC 2009). The current cost of electricity generated with different technologies is shown in Table 4.3.

The model discussed here considers cost escalation for each power generation technology over the projected period on the basis of an increase in overnight costs. These assumptions have been checked against other costing studies for consistency (van den Broek et al. 2009; Pauschert 2009; IEA 2011). The cost projections for the future also incorporate current estimates of new technology such as carbon capture

Table 4.3 Cost of electricity from various fuel sources in India

(2010 Constant prices)	Capital cost (million$/MW)	Energy costs (cents/kWh)
Large hydro	1.52	4.65
Coal (average across imported and domestic coal)	1.04	4.69
Natural gas	0.73	8.81
Nuclear (average across imported and domestic nuclear)	1.50	5.83
Wind	1.46	9.79
Small hydropower	1.38	4.94
Biomass	1.23	7.71
Solar photovoltaic	1.44	8.77
Concentrated solar power	2.58	20.98

Data source: Calculated for each technology using Central Electricity Regulatory Commission *(Terms and Conditions of Tariff)* Regulations
Conversion to $ from Rs. at 2009 rate of 48.04

and sequestration (CCS) or integrated gasification combined cycle (IGCC) for gas- and coal-based generation. The trends of electricity costs might well change in the future, and can be modified accordingly. Estimates of cost escalation given in the literature show a wide range of variation and have assumed to both increase and decrease for a time period of the next 40 years.

Similarly, the value of the discount rate to be applied for such an analysis done for the developing country also varies significantly in the literature. The WGIII report in IPCC AR4 (IPCC AR4, WG-III, 2013) provides a detailed discussion of the discount rates to be used for developed and developing countries. It recommends that though discount rates of between 4 and 6% may be justified for long-term mitigation projects in the developed countries, in developing countries usually the rates used may be higher (10–12%). Some studies use a discount rate of 8% (Mallah and Bansal 2010). There are recommendations that suggest that for longer time periods (more than 30 years), discount rates should progressively reduce (OXERA 2002). However, this further complicates the problem and for the sake of simplicity a single discount rate of 10% is used in this analysis. However in estimating relative costs or additional costs as a percentage of baseline costs, the discount rate is not a serious issue.

4.3   Power Sector Model: Methodology

The model presented here is an expanded version of a typical generation expansion model and provides a least-cost fuel mix for a power sector subject to multiple constraints. The power sector model is illustrated in Fig. 4.1.

The problem is one of linear optimization, solved using the GAMS (General Algebraic Modeling System) platform, wherein the total pooled cost of electricity— both fixed cost and variable cost—for the time period 2010–2050 is minimized. The demand curve for electricity is constructed by the model based on input requirements specified by the user as explained above.

The objective function is specified as shown in Eq. (4.1):

$$\text{OBJ} = \text{Min}\left(\sum_F \sum_T \text{FC}_{F,T} + \sum_F \sum_T \text{VC}_{F,T}\right) \tag{4.1}$$

where OBJ is the objective function and $\text{FC}_{F,T}$ is the total pooled fixed cost (capital investment) for each power generation technology F and each year T—in Rs./MW discounted to 2010 values. T varies from 2010 to 2050. F = (coal, hydro, nuclear, natural gas, diesel, wind, biomass, small hydro power, solar photovoltaic, concentrated solar power, municipal solid waste). $\text{VC}_{F,T}$ is the total variable cost of power generation technology and each year—in Rs./kWh discounted to 2010 values. The cost is minimized subject to constraints specified below.

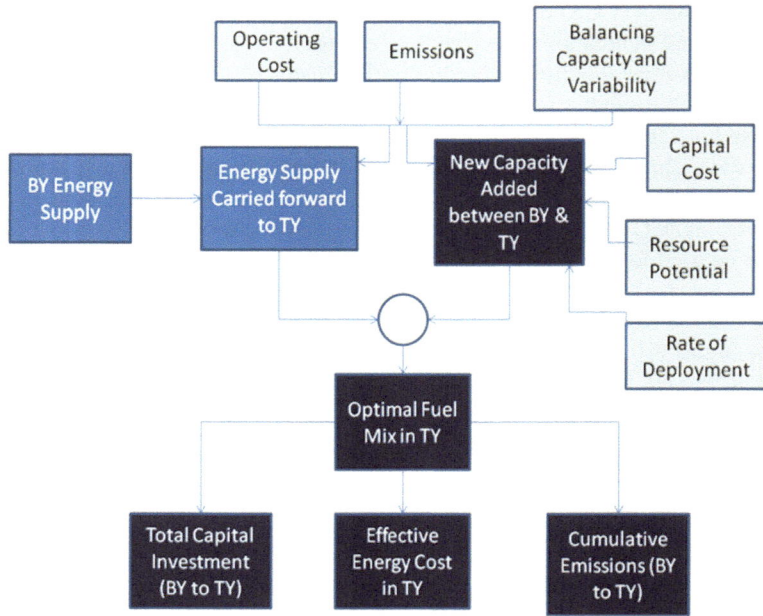

Fig. 4.1 Power sector model—overview of methodology; *BY* base year, *TY* target year

The total supply of electricity has to match the total demand in each time period T. Equations (4.2) and (4.3) are the basic equations used to generate values for the main variables—installed capacity for each power generation technology and actual electricity generation from the same:

$$\sum_F ES_{F,T} = D_T - \sum_F EG_{F,T} \tag{4.2}$$

$$\sum_F IC_{F,T} = \sum_F ES_{F,T} \times 8760 \times PLF_{F,T} \tag{4.3}$$

where $ES_{F,T}$ is the electricity supplied by each technology in year T. D_T is the demand for electricity in year T. $EG_{F,T}$ is the electricity supplied by existing plants in year T. $IC_{F,T}$ is the installed capacity for each technology in year T. $PLF_{F,T}$ is the plant load factor or capacity factor for each technology in year T.

The cumulative emission constraint (with the actual budget input as a constant fraction of the total budget available to the country, which may be different for each scenario that is generated) is given as a hard constraint that cannot be violated shown in Eqs. (4.4) and (4.5):

$$AEM_T = \sum_F ES_{F,T} \times EF_F \tag{4.4}$$

$$\sum_{T} \text{AEM}_T \leq \text{CB} \tag{4.5}$$

where AEM_T denotes the annual emissions from the power sector in year T. EF_F is the emission factor for each power generation technology (in tC/kWh). CB is the carbon budget (cumulative emissions) available to the country between 2010 and 2050.

Resource constraints are input as simple limits on the total generation or installed capacity for each fuel source. For each fuel source there is a limit on the total cumulative potential available in the country. Also, for each fuel source, there is a limit applied to the new power generation capacity that can be reasonably commissioned in each year and the capacity that can be decommissioned. For example it is assumed that no more than 30–50% (varying based on the type of power generation technology) of current installed capacity can be commissioned anew in each year. The model also has to obey constraints for base and peak load. It is assumed here that total base load capacity at any given time cannot fall below 30% of the total generation capacity. The operationalization of these constraints is shown in Eqs. (4.6), (4.7), and (4.8):

$$\sum_{T} \text{IC}_F \leq \text{AP}_F \tag{4.6}$$

$$\begin{aligned} \text{IC}_{F,T+1} &\leq a \times \text{IC}_{F,T} \text{ and} \\ \text{IC}_{F,T+1} &\geq b \times \text{IC}_{F,T} \end{aligned} \tag{4.7}$$

$$\text{BES}_T \geq 0.3 \times \text{ES}_T \tag{4.8}$$

where AP_F is the cumulative potential for each fuel. BES_T is the total base load contribution to electricity generation in year (T). "a" and "b" are coefficients used to express the limits within which new installed capacity can be added or decommissioned.

Additional constraints have also been added in the model to account for other limits that operate on power generation. For example, the average plant life considered for existing as well as additional thermal power plants is 25 years. If any plant is decommissioned before its entire lifetime, then it would follow that the cost of electricity from that plant is higher than the values considered as inputs to the model. This additional cost in case of all plants is not accounted for in the cost calculations. However an additional constraint is applied specifically for coal-based generation, wherein the emissions from the plant added in the first year are committed for the subsequent 25 years. The constraint is operationalized by disallowing decommissioning of newly added coal plants for the first 25 years. However existing coal and gas plants will be decommissioned as per scheduled lifetime estimates before this 25-year time period.

It should also be noted that the model provides the fuel mix transitions based on a carbon budget available to the country between 2010 and 2050. The carbon budget available beyond 2050 is not accounted for in the model. In addition to these constraints there is also a constraint on balancing that is added to the model. Renewable energy sources introduce a certain amount of variability in the system. A detailed analysis of which particular plants can be used to balance this variability would require an analysis of load dispatch which is not the purpose of this exercise. However, renewable energy technologies especially solar and wind energy technologies would require one of the two things—(i) the presence of a considerable amount of storage capacity to balance the variability which in some months can be very high even on an hourly basis or (ii) the presence of some other balancing fuel technologies in the system that can be ramped up or down at fast rates to balance the variability. For the sake of simplicity in this analysis the balancing capacity in the system is represented by the capacity factors of conventional power plants. For each technology there is lowest capacity factor that a plant of a particular technology can operate at without losing efficiency. In this analysis, hydro and gas plants are considered to be the potential balancing plants. However, this means that the total capacity hydro and natural gas plants in the system at any given point in time have is also a factor of the amount of solar and wind energy plants sought to be installed in the system. The balance of the renewable energy variability is made up by storage. Equation (4.9) summarizes this balancing requirement introduced in the model:

$$\underset{F=\text{Solar, Wind}}{\text{IC}}_{F,T} \leq \underset{F=\text{Hydro, Gas}}{\text{IC}}_{F,T} + S_T \tag{4.9}$$

where S_T is the total amount of storage capacity required in the system. The key variables used in the model and the key constraints are summarized in Table 4.4.

4.4   Illustration of Power Sector Model Through Scenarios

The model can be run for various scenarios of electricity demand and various constraints. The results presented here are for the scenarios outlined in Table 4.5. In each scenario a successively more stringent constraint on cumulative emissions for India is considered.

The first scenario provides a baseline for the cost incurred for power generation in the absence of a constraint on emissions. For each subsequent scenario the fuel mix is further constrained by a budget. The global carbon budgets are based on estimates provided by the Fifth Assessment Report of the IPCC for restricting temperature rise to 2 °C. If we want at least a 50% probability of restricting temperature rise to below 2 °C, then the world can emit no more than 67 GtC between 2012 and 2100. Scenarios 2 and 3 are based on this global emission constraint. It is assumed that 70% of this cumulative emission quota is available between 2012 and 2050. In Scenario 2 it is assumed that India can access its complete per capita entitlements (17% of total cumulative emissions as India has 17% of the total global population). In Scenario 3, only half of this is available to India. This takes into consideration the

Table 4.4 Key variables and constraints

Key variables	Description	
Energy supply	11 fuel technologies over 20 years—Hydro, coal, diesel, gas, nuclear, wind, SHP, biomass, solar PV, CSP, MSW	$ES_{F,T}$, $IC_{F,T}$
Emissions	Operating emissions for four technologies over 20 years	$EM_{F,T}$
Capital cost	Cost of new capacity of each type over 20 years	$CC_{F,T}$
Operating/variable cost	Operating cost of each technology over their lifetimes	VC_f
Key constraints	Description	Equations
Resource potential	For all kinds of plants	$\Sigma IC_{F,T} < A$
Balancing capacity in the system	(i) Represented by capacity factor. Lowest capacity factor that can be achieved without reducing efficiency	$CF_{F,T} > B$
	(ii) By the amount of hydro and gas (fast ramping) capacity and storage needed to manage variability	$IC_{F,T}$ (solar, wind) $= IC_{F,T}$ (gas, hydro) $+ S_T$
Base load requirements	Assumed to be 30% of total demand	$BES(t) \geq 0.3 \times ES(t)$
Rates of construction and commissioning	Rate at which plants of each technology can be added—longer for nuclear, large hydro	$IC(f,t+1) \leq a \times IC(f,t)$
Rates of decommissioning	Rates at which plants of each technology can be decommissioned	$IC(f,t+1) \geq b \times IC(f,t)$

possibility of developed countries not undertaking their share of mitigation actions, leaving a large burden to be borne by developing countries such as India. Similarly for a higher probability of 67% of restricting temperature rise to below 2 °C, the emission quota is 477 GtC between 2012 and 2100. Scenarios 4 and 5 are based on this emission constraint on global emissions.

If India is to achieve its target of 2500 kWh/person/year in 2030 and 3500 kWh/person/year in 2050 under these constraints through a least-cost pathway, the cost to the economy is shown in Table 4.6. Additional cost in this table refers to the total costs incurred for all new power generation capacity added between 2010 and 2050.

The total cost incurred in adding new power generation capacity under Scenario 5 is about 107% higher than estimated under Scenario 1 for the time period 2012–2030 and about 111% higher for the time period 2012–2050. The carbon constraint is the highest in Scenario 5. It can be seen from Table 4.6 that the costs progressively increase with a tighter carbon constraint for the power sector. According to the 12th 5-year plan prepared by the Indian Government, the total investments in the power sector required between 2012 and 2017 were estimated to be about Rs. 2 trillion. In reality slightly less than half of this projected requirement was actually invested in the sector. In Scenario 1, the investments required amount to about Rs. 0.75 trillion annually between 2012 and 2030, which approximately matches the estimates in the 12th 5-year plan. In Scenario 5 however, the required

Table 4.5 Five scenarios for determining fuel mix for India—base year—2004, target year—2050

Scenarios for carbon budgets		Total budget for India (GtC)— 2012–2050	Total budget for the power sector (GtC)	Budget for the power sector after accounting for existing power plants (GtC)
Scenario 1: No global restriction on emissions	No restriction on emissions between 2012 and 2050	–	–	–
Scenario 2: Low global and national restrictions	Global carbon constraint—697 GtC between 2012 and 2100 (assume 70% of it available till 2050—488 GtC); India gets full per capita entitlement—17%	83	33	29
Scenario 3: Low global but high national restrictions	Global carbon constraint—697 GtC between 2012 and 2100 (assume 70% of it available till 2050—488 GtC); India gets only half of its full entitlement (as developed countries use more space)	41	21	17
Scenario 4: High global and low national restrictions	Global carbon constraint—477 GtC between 2012 and 2100 (assume 70% of it available till 2050—334 GtC); India gets full per capita entitlement—17%	57	23	19
Scenario 5: High global and national restrictions	Global carbon constraint—477 GtC between 2012 and 2100 (assume 70% of it available till 2050—334 GtC); India gets only half of its full entitlement (as developed countries use more space)	28	14	10

investment is about Rs. 1.6 trillion, an increase of almost 100%. The fuel mix transitions for Scenarios 1, 3, and 5 for an energy requirement of 2500 kWh/person/year in 2030 and 3500 kWh/person/year in 2050 are shown in Fig. 4.2.

As seen from Fig. 4.2, coal-based power generation is severely restricted with each additional restriction on cumulative carbon budget. In each of the three scenarios a large part of the electricity demand in the final years is supplied by rapid addition of renewable energy technologies, especially solar photovoltaic. This happens for two reasons—(i) cost escalation estimates predict that the real costs for

Table 4.6 Additional costs for new power generation between 2012 and 2050 for each scenario[a]

Scenarios	Cumulative carbon constraint between 2012 and 2050 (GtC)	Cumulative capital cost from 2012 to 2030 (Rs. trillion)	% Increase as compared to Scenario 1	Cumulative capital cost from 2012 to 2050 (Rs. trillion)	% Increase as compared to Scenario 1
1	No constraint	13.5		25	
2	29	13.7	1%	37	51%
3	17	16.9	25%	45	81%
4	19	16.2	19%	44	76%
5	10	28.1	107%	53	111%

Conditions—10% discounting, 5% escalation
[a]Table shows the total cost incurred for all new power generation undertaken to meet energy requirements between 2012 and 2050

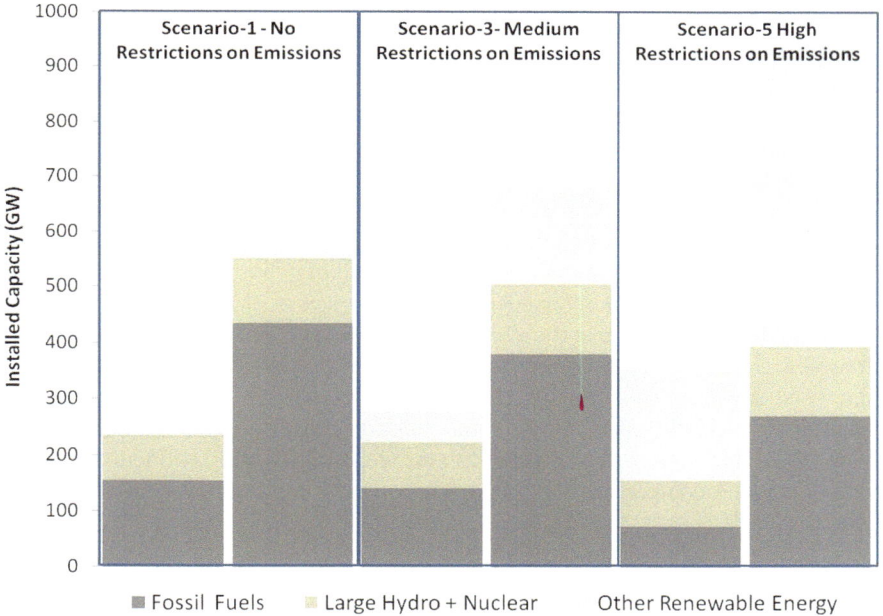

Fig. 4.2 Fuel mix transitions for Scenarios 1, 3, and 5

solar technology will reduce in the coming years before saturating and increasing later with a commensurate increase in material costs and other allied costs. However this reduction enables solar technology to achieve grid parity by about 2020; (ii) there is no limit specified for the maximum amount of solar energy generation that can be installed in the country as there is currently no methodology to establish this reliably. For all other fuel technologies however, a maximum resource limit is specified beyond which no more capacity of that particular fuel type can be added.

Table 4.7 Some illustrative results for key variables in Scenarios 1 and 5

Variable	Variable description	Lower limit	Upper limit	Value in Scenario 1	Value in Scenario 5
IC (hydro, 2020)	Installed capacity of hydropower in 2020	None	150 GW	77 GW	77 GW
IC (coal, 2030)	Installed capacity of coal-based power in 2030	None	600 GW	379 GW	101 GW
IC (wind, 2030)	Installed capacity of wind-based power in 2030	None	300 GW	13 GW	127 GW
CEM	Cumulative emissions between 2009 and 2050	None	Specified by constraint for each scenario	42 GtC	10 GtC

This leads to the selection of solar technology beyond a particular time as it becomes more cost effective and the most abundantly available resource. It should be noted however that it may not be possible to achieve the levels of generation prescribed in the model with solar technology. In the absence of any breakthrough in other technology and high constraints on emissions, this would mean that the total energy demand in 2050 may not be met. Table 4.7 shows illustrative results for some key variables for Scenarios 1 and 5.

From the relative additional burden in increasingly constrained budget scenarios it is clear that the availability of adequate carbon space is a key constraint as one may have intuitively expected at the very outset. However it is difficult to obtain precise absolute estimate of costs as they are sensitively dependent on the cost scenarios that are used in the model. It is entirely possible that the cost projections used in this model may in reality change in the future. In such a case, the additional burden in the various budget scenarios as compared to the unrestricted or full entitlement case will also change. A comparison of costs based on two different values of discount rate for each scenario is shown in Fig. 4.3.

The results shown in Fig. 4.3 illustrate the sensitivity of the model to the assumptions made in the model. All models are sensitive to such assumptions. The results of the model therefore should not be taken as predictive but as indicative of the implications of certain policy decisions. In addition to these considerations it is interesting to examine the emission profile from each of these scenarios, which, for ease of comparison, have been plotted together in Fig. 4.4.

The shape of the profiles indicates that as the budget is increasingly constrained, the peak of the emission profile is reduced with the peak also moving back to earlier years. In some of the constrained scenarios, the annual emission rises up in the last few years since some coal capacity is added towards the end of the time period to make up a minimum base load requirement of 30%. In Scenario 1, the smooth increase and then plateau of emissions correspond to the steady addition of fossil fuels in the absence of any constraint on emissions. On the other hand, in Scenario

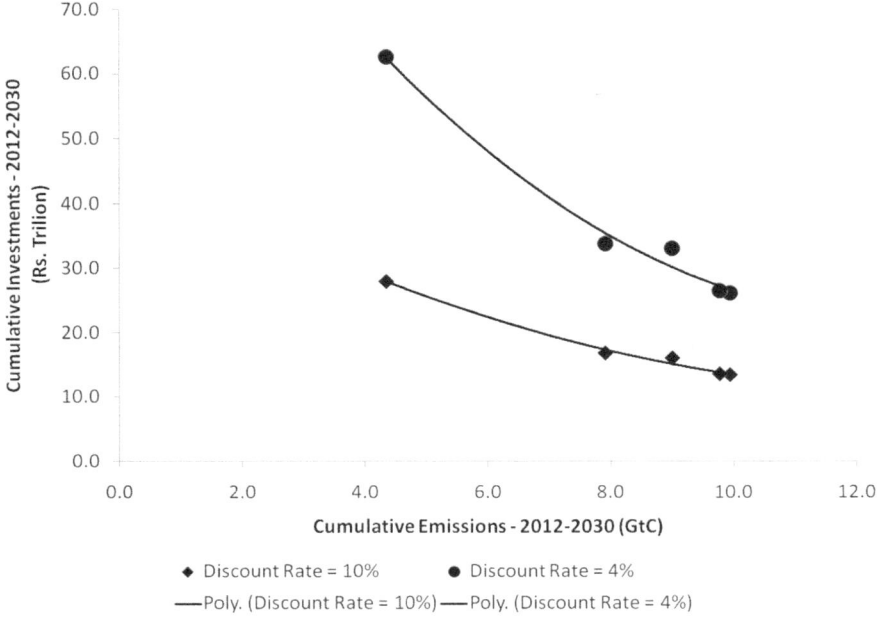

Fig. 4.3 Investments incurred in each scenario for two values of discount rate—4 and 10%

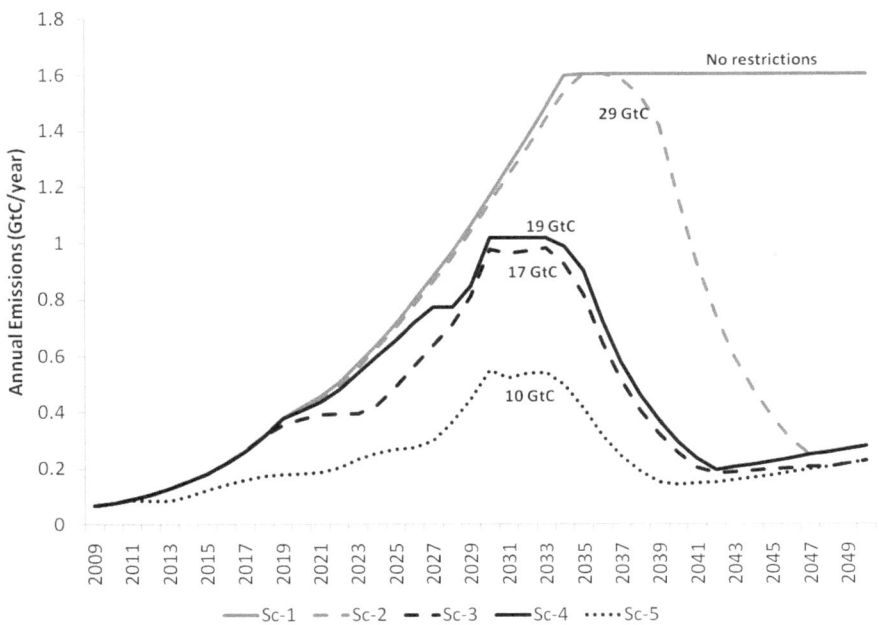

Fig. 4.4 Annual emissions for all five scenarios

5, the value of peak emissions at about 0.45 GtC/year corresponds to the fact that there is very little new fossil fuel capacity added. The emission intensity of electricity in 2009 is 0.27 kg_{CO_2}/kWh. In all the scenarios this value increases in the short term, i.e., till 2030 when more coal-based power plants are being added, and reduces in Scenarios 3 and 5 by 2050. The results also point to the fact that even with increasing emission intensity values in the short term it is possible to stay within a cumulative emission constraint. Mitigation targets that are proposed in the form of emission intensities for the short term may therefore unnecessarily restrict countries from adopting optimal energy pathways. Results show therefore that a cumulative emission constraint instead of an annual constraint on emissions or a restriction on the value of emission intensity can provide the flexibility to decide when and when not to deploy certain technologies.

4.5 Conclusion

India's per capita electricity consumption in 2010 was 705 kWh/person/year. As compared to the per capita electricity consumption in developed countries, this is significantly low (USA—11,920 kWh/person/year, France—7023 kWh/person/year, the UK—5467 kWh/person/year, Portugal—4477 kWh/person/year) (World Bank 2013). The results presented here clearly show that even for a normative goal of 2500 kWh/person/year in 2035 and 3500 kWh/person/year in 2050, the investments required in the power sector are significantly high. Considerations of global sustainability operationalized in the model through a cumulative constraint on emissions add further to this burden. Model results show that a 30% reduction in the available carbon space increases the costs by 25% till 2030 and about 80% till 2050. As the constraint increases, either due to a stringent constraint on global emissions or due to India's inability to access its entire entitlement to carbon space, this cost increases further.

Cheaper sources of energy are completely accessible only in Scenario 1, where there are no restrictions on emissions. New coal-based power capacity that can be added reduces by 80% under Scenario 2 and reduces by 90% under Scenario 3. This essentially means that, under a mitigation regime, wherein the future carbon space available to India is severely restricted, only about 85 GW of new coal-based capacity may be added. In the 12th 5 Year Plan Report of the Government of India (GoI 2012), it is estimated that about 62 GW of coal-based capacity would be required to be added in the next 5 years itself. This also has serious implications for the availability of base load power required for industry. The only other technology that may be able to provide base load power in this scenario is nuclear energy, which is accompanied by a significant increase in costs. A tighter budget restriction would also leave little flexibility for a smooth transition to a low carbon economy. It is important to note that India would require both a higher carbon budget for the future and an increase in the level of renewable energy capacity as a part of the total fuel mix. A higher budget will not eliminate the necessity to increase the share of renewable energy, especially if the country wishes to achieve higher levels of per capita energy use in the future.

The peaking year for annual emissions is 2035 for Scenario 1, after which emissions stabilize as no new coal-based capacity is added. The level at which emissions peak reduces for each subsequent scenario. The peak emissions in Scenario 3 are about 40% lower as compared to Scenario 1 and in Scenario 5 they are about 98% lower. The year of peaking also shifts to earlier years. This is another indicator of the reduction in flexibility available to the power sector in undertaking a transition to a low carbon pathway.

A key contribution of this work is the presentation of a method of evaluating national policy options in the power sector which is explicitly linked to global environmental concerns. It is of crucial importance to developing countries, and especially large developing economies, to connect global mitigation efforts to the implications of the same for the development of their national energy infrastructure. The investment estimated as a result of this fuel mix can be used as an input to the input-output model for scenarios that evaluate the impact of higher deployment of renewable energy technologies in the power sector on other sectors of the economy and on key developmental indicators. This module as part of the integrated modeling framework provides a way in which the specific characteristics of fuel technologies and the energy system as a whole can be included in an integrated energy-economy-environment analysis. The integration is discussed further in Chap. 6. The next chapter discusses the third component of the integrated modeling framework which is input-output analysis.

References

Central Electricity Authority, India. (2009). All India Electricity Statistics: General Review 2009, New Delhi.

Central Electricity Authority. (2011). *All India Electricity Statistics 2011: General review 2011.* New Delhi: Central Electricity Authority.

Central Electricity Regulatory Commission. (2009). Central Electricity Regulatory Commission (Terms and Conditions of Tariff) Regulations. New Delhi.

GoI. (2006). Government of India, Office of the Principal Scientific Adviser and the Energy and Resource Institute. National Energy Map for India: Technology Vision 2030. New Delhi: TERI Press.

GoI. (2012). Planning Commission, Government of India. Report of the working group on power for 12th plan. New Delhi, Ministry of Power.

International Energy Agency. (2011). World Energy Outlook 2011 Report, Assumed investment costs, operation and maintenance costs and efficiencies for power generation in the New Policies and 450 Scenarios. Retrieved April 23, 2015, from http://www.worldenergyoutlook.org/weomodel/investmentcosts/

Mallah, S., & Bansal, N. K. (2010). Allocation of energy resources for power generation in India: Business as usual and energy efficiency. *Energy Policy, 38*(2), 1059–1066.

OXERA, O. E. (2002). *A report to the DTI and the DTLR Regional Renewable Energy Assessments.* Oxford: Oxera Environmental.

Pauschert, D. (2009). Study of equipment prices in the power sector. ESMAP Technical Paper 122/09. Energy Sector Management Assistance Program. Washington, DC: The World Bank Group.

van den Broek, M., et al. (2009). Effects of technological learning on future cost and performance of power plants with CO_2 capture. *Progress in Energy and Combustion Science, 35*(6), 457–480.

World Bank, World Development Indicators. (2012). *Per capita electricity consumption.* Retrieved from https://data.worldbank.org/country/india?view=chart.

World Bank, World Development Indicators. (2013). *Per capita electricity consumption.* Retrieved from https://data.worldbank.org/country/portugal?view=chart.

Yeh, S., & Rubin, E. S. (2012). A review of uncertainties in technology experience curves. *Energy Economics, 34*(3), 762–771.

Chapter 5
Analysis of Energy-Economy Linkages Using a Social Accounting Matrix

Abstract The third component of the integrated modeling framework, discussed in this chapter, is the input-output analysis, which provides insights into key economic variables that are affected due to different energy policy decisions. In the climate change debate, developing countries have always voiced concerns over taking mitigation action as they believe that this will affect the pace and way in which they can develop and address the issues of poverty alleviation within their countries. On the other hand developed countries have also been averse to undertaking mitigation action citing a loss of competitive advantage if they are to be the sole contributors to the mitigation effort. In both arguments, the trade-offs associated with the development versus environment debate are implicit. However, an analysis of the nature of the trade-offs or their quantification has been done in very few studies. For India, most modeling studies are undertaken with a view to produce projections for the future which will be useful in determining India's energy requirements, emissions, and economic growth. Such projections, however useful, cannot be taken as accurate projections of the future as they are but one possible representation out of a range of possible futures. Every change in policy leads to an impact, large or small, on all the sectors of the economy, and this leads to an impact on energy and emissions. The best that is possible therefore is to evaluate a range of scenarios for a range of policy alternatives and, instead of focusing on providing a single predictive result, focus on evaluating the interlinkages in the economy and the potential impacts of certain decisions on key economic and energy parameters. The work presented in this chapter undertakes such an analysis of "what ifs" using an extended input-output matrix, i.e., a social accounting matrix (SAM).

Keywords Social accounting matrix · Input-output models · Energy equity · Energy-economy linkages · Structural decomposition

The third component of the integrated modeling framework is the input-output analysis, which provides insights into key economic variables that are affected due to different energy policy decisions. In the climate change debate, developing countries have always voiced concerns over taking mitigation action as they believe

T. Kanitkar, *An Integrated Framework for Energy-Economy-Emissions Modeling*, SpringerBriefs in Environmental Science, https://doi.org/10.1007/978-3-030-18263-2_5

that this will affect the pace and way in which they can develop and address the issues of poverty alleviation within their countries. On the other hand developed countries have also been averse to undertaking mitigation action citing a loss of competitive advantage if they are to be the sole contributors to the mitigation effort. In both arguments, the trade-offs associated with the development versus environment debate are implicit. However, an analysis of the nature of the trade-offs or their quantification has been done in very few studies. Studies such as the Stern Review (Stern 2006) assume an aggregated damage function which calculates the impact of mitigation action on adaptation requirements and consequently on economic growth. The implications of climate change mitigation for economic growth are an assumption in these studies rather than outcomes of modeling. Single-country studies, such as the study by Parikh and Ghosh (2009) for India, for example, do consider the impacts of mitigation actions (deploying renewable energy technologies) on overall economic growth and also estimate overall impacts on income distribution. The impacts of decisions about the pattern of economic growth and electricity generation on specific household classes however are not reported in these studies. On the other hand, most studies that are done to evaluate the impacts of renewable energy technology on these parameters (viz., income, equality, and employment) are restricted to evaluating local impacts of introducing renewable energy technologies (Myles 1998; del Río and Mercedes 2009; Apergis and Payne 2014).

For India, most modeling studies are undertaken with a view to produce projections for the future which will be useful in determining India's energy requirements, emissions, and economic growth. Such projections, however useful, cannot be taken as accurate projections of the future as they are but one possible representation out of a range of possible futures. Every change in policy leads to an impact, large or small, on all the sectors of the economy, and this leads to an impact on energy and emissions. The best that is possible therefore is to evaluate a range of scenarios for a range of policy alternatives and, instead of focusing on providing a single predictive result, focus on evaluating the interlinkages in the economy and the potential impacts of certain decisions on key economic and energy parameters. For example, instead of focusing on whether the per capita emissions for India in 2030 will be 3.5 tCO_2 or 5.1 tCO_2, it may be more useful to focus on the implications of these two emission values for key economic variables such as incomes, income distributions, economic growth, employment, and impact of changes in these parameters in turn on energy requirements and emissions.

5.1 Introduction

The work presented in this chapter undertakes such an analysis of "what ifs" using an extended input-output matrix, i.e., a social accounting matrix (SAM). A social accounting matrix—SAM—is a representation of the macroeconomic accounts of a socioeconomic system, which captures the transactions and transfers between all economic agents in the system (Pyatt and Round 1985; Pyatt 1988; Reinert and

Roland-Holst 1997). The SAM itself is not a model. It is a static representation of a set of macro data for an economy for a defined time period. However, the SAM can be used as a basis for creating a dynamic model which evaluates different scenarios and outcomes for the future. Pyatt, Thorbecke, and others have used the SAM to address issues such as poverty and income distribution in developing countries in the 1970s. The models used by them were SAM-based multiplier models in which column shares (column coefficients) are computed from a SAM in order to compute matrix multipliers. A detailed explanation of this methodology is given in Round (2003). A review of other studies that use a social accounting matrix as the basis is given in Chap. 2.

The SAM is used in this analysis not as a basis for a CGE model but as a basis for simulating scenarios for the future. The multiplier decomposition method detailed by Round (2003) is used for this purpose. A detailed explanation of the methodology used in this analysis is given in the next section. The analysis presented in this analysis provides a methodology to evaluate the impact of high deployment of renewable energy technologies on economic growth, incomes, and equality. It is possible using this method to understand the nature of the trade-off, i.e., whether inequality reduces or increases with higher deployment of renewable energy and also to quantify the nature of the trade-off. Although the quantification is subject to many assumptions of costs, efficiency, etc. it is possible to vary these assumptions in order to arrive at a range of possible outcomes which relate the energy and economic decision-making processes. The next section gives a brief introduction to social accounting matrices and their use in energy analysis. The subsequent section explains the methodology used in this analysis. The scenarios constructed using this modeling framework and the results for four basic indicative scenarios are then discussed in later sections.

5.2 Introduction to SAM

A SAM is a comprehensive accounting framework within which the full circular flow of income is captured. In a SAM all the transactions in the economy are presented in the form of a matrix. Each row of this matrix gives receipts of an account and each column gives the expenditure. The total of each row i is supposed to be equal to the total of each corresponding column j (where $i = j$). An entry in row i and column j represents the receipts of account i from account j. Table 5.1 shows a schematic of a typical SAM.

The basic structure of SAM is based on the following transactions and transfers in the economy; production requires intermediate goods and primary factors of production, viz., labor and capital (in some social accounting matrices, land is also included as a factor of production). These factor endowments are contributed by the institutions, viz., households, firms, and government, who in turn receive factor payment—value added. Apart from value added, institutions get income from other sources, such as transfers from the government and from the rest of the world.

Table 5.1 Schematic of a typical SAM

	Commodities	Activities	Factors	Households	Other accounts	Total
Commodities		Intermediate consumption		Household final consumption expenditure	Other final demands	Total demand for products
Activities	Domestic supplies					Total activity outputs
Factors		Value added			Factor income from abroad	Total factor income receipts
Households			Factor income to household	Inter-household transfers	Nonfactor income receipts	Total household incomes
Other accounts	Imports	Indirect taxes	Other factor payments	Savings		Total exogenous receipts
Total	Total supply of products	Total activity outputs	Total factor income payments	Total household outlays	Total exogenous payments	

Source: Round (2003)

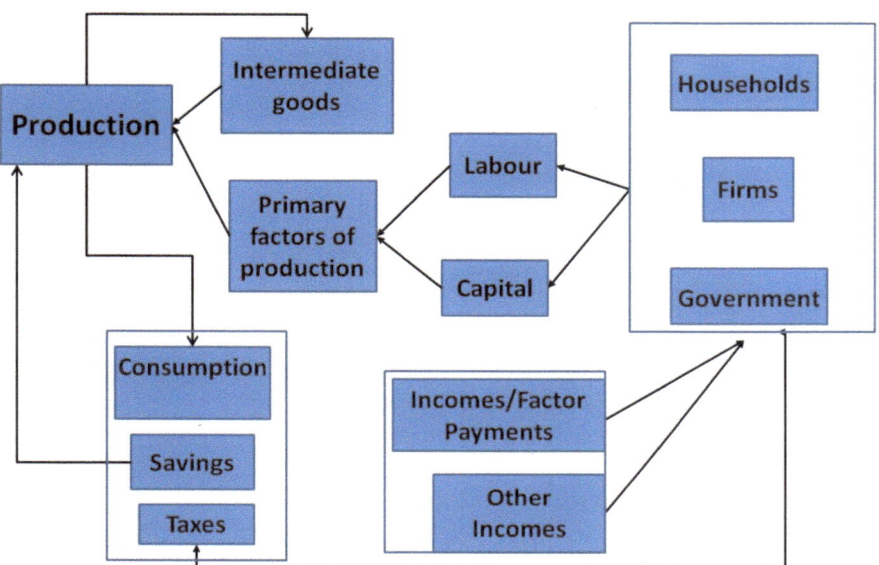

Fig. 5.1 Circular flow of income in a social accounting matrix

Income is spent as the consumption expenditure on goods and services and for payment of taxes and the rest is saved for the future. Total supply in the economy has to be matched by the demand made by the institutions and capital formation—purchase of investment goods. In the SAM, extra breakdown of the household sector is done to reflect the role of different household classes in the economy. A schematic of the circular flow of income is shown in Fig. 5.1.

Each row of the SAM represents a linear relationship between the inputs to a sector and the outputs. The equations are detailed as follows:

1. Gross output = Total cost of production:

$$\sum_{j=1}^{n} Y_{ij} = \sum_{i=1}^{n} \left[RM_{ij} + VA_{ij} + TIG_{ij} \right] \tag{5.1}$$

where Y_{ij} is the gross output of the sector i distributed across each sector j. RM_{ij} is the total purchase of raw material from sector i for producing one unit of output in sector j. VA_{ij} is the total value added by labor and capital for production of one unit in sector j. TIG_{ij} is the taxes paid on the purchase of intermediate goods by sector j.

2. Aggregate demand = Aggregate supply:

$$\sum_{j=1}^{n} \left[RM_{ij} + HC_{ij} + GC_{ij} + CF_{ij} + EX_{ij} \right] = \sum_{i=1}^{n} \left[Y_{ij} + IM_{ij} \right] \tag{5.2}$$

where HC_{ij} is the final household consumption of all household classes of output from sector I. GC_{ij} is the final consumption by government of output from sector i. CF_{ij} is the gross capital formation in sector i. EX_{ii} is the amount of output from sector i that is exported. IM_{ji} is the amount of demand for products from sector j that are imported.

3. Total factor income = Total factor endowment:

$$\sum_{j=1}^{n} \left[VA_{ij} + IA_{ij} \right] = \sum_{i=1}^{n} \left[H_{ij} + PPr_{ij} + PPu_{ij} + GI_{ij} + DC_{ij} \right] \tag{5.3}$$

where IA_{ij} is the factor income earned from abroad by capital and labor. H_{ij} is the total endowment (income) of households. PPr_{ij} is the total operating profit of private corporations. PPu_{ij} is the total operating surplus of public enterprises. GI_{ij} is the total income of the government from public enterprises. DC_{ij} is the total depreciation.

4. Total household income = Total household expenditure + Total household saving:

$$\sum_{j=1}^{n} \left[H_{ij} + G_{ij} + HA_{ij} \right] = \sum_{i=1}^{n} \left[HC_{ij} + DTH_{ij} + ITH_{ij} + SH_{ij} \right] \tag{5.4}$$

where G_{ij} is the direct monetary transfer from the government to households. HA_{ij} is the monetary transfer received by household classes from abroad. DTH_{ij} is the tax paid by household classes to the government. ITH_{ij} is the indirect tax paid by household classes in purchase of commodities. SH_{ij} is the total savings of household classes.

5. Income of private corporations = Use of income of private corporations:

$$\sum_{j=1}^{n} \left[PPr_{ij} + IPr_{ij} \right] = \sum_{i=1}^{n} \left[TPr_{ij} + SPr_{ij} \right] \tag{5.5}$$

where IPr_{ij} is the interest on debt paid by private corporations to the government. TPr_{ij} is the corporate taxes paid by private corporations to the government. SPr_{ij} is the savings account of private corporations.

6. Income of public enterprises = Use of income of public enterprises:

$$\sum_{j=1}^{n} PPu_{ij} = \sum_{i=1}^{n} SPu_{ij} \tag{5.6}$$

where SPu_{ij} is the savings account of public enterprises.

7. Total government earnings = Total government expenditure:

$$\sum_{j=1}^{n} \left[\text{GI}_{ij} + \text{DTH}_{ij} + T\text{Pr} + \text{Tin}_{ij} + \text{GTA}_{ij} \right]$$

$$= \sum_{i=1}^{n} \text{GC}_{ij} + G_{ij} + I\text{Pr}_{ij} + \text{TG}_{ij} + \text{SG}_{ij} \tag{5.7}$$

where Tin_{ij} is the total indirect tax accruing to the government. GTA_{ij} is the net capital transfer to the government from abroad. TG_{ij} is the tax paid by government on purchases of goods and services. SG_{ij} is the total savings account of the government.

8. Total indirect taxes paid = Total indirect taxes received:

$$\sum_{j=1}^{n} \left[\text{TIG}_{ij} + \text{ITH}_{ij} + \text{TG}_{ij} + \text{TC}_{ij} + \text{TE}_{ij} \right] = \sum_{i=1}^{n} \text{Tin}_{ij} \tag{5.8}$$

where TC_{ij} is the tax paid on investment goods. TE_{ij} is the tax paid in exports.

9. Gross savings in the economy = Aggregate investment in the economy:

$$\sum_{j=1}^{n} \left[D_{ij} + \text{SH}_{ij} + \text{SPr}_{ij} + \text{SPu}_{ij} + \text{SG}_{ij} + \text{SA}_{ij} \right] = \sum_{i=1}^{n} \left[\text{CF}_{ij} + \text{TC}_{ij} \right] \tag{5.9}$$

where SA_i is the total foreign savings.

10. Foreign exchange payments = Foreign exchange receipts:

$$\sum_{j=1}^{n} I_{ij} = \sum_{i=1}^{n} \left[E_{ij} + \text{IA}_{ij} + \text{HA}_{ij} + \text{GA}_{ij} + \text{TE}_{ij} + \text{SA}_{ij} \right] \tag{5.10}$$

5.2.1 Multiplier Analysis Using SAM

From the SAM it is possible to calculate the effect of an injection into any endogenous account (vector f) on the income or output of all endogenous accounts using multiplier analysis. Exogenous injections reflect government consumption, investment goods demand, and exports. The multiplier matrix is derived from the matrix of expenditure propensities S, which is obtained by dividing each entry in the endogenous accounts by its respective column total. The vector of total income of endogenous accounts $\bar{x}$ can be expressed as

$$\bar{x} = S\bar{x} + \bar{f} \tag{5.11}$$

where $\bar{x}$ is an $(n \times 1)$ column vector of total output of all n endogenous accounts, S is the $(r \times n)$ matrix of average expenditure propensities/SAM coefficients, and $\bar{f}$ is an $(n \times 1)$ column vector of exogenous injections. From this expression it follows that

$$\bar{x} = I - S^{-1}\bar{f} = M_A\bar{f} \tag{5.12}$$

where M_A is the $(n \times n)$ matrix of accounting multipliers. The matrix M_A is similar to the Leontief matrix of inter-industry coefficients in an input-output analysis. A matrix Z can be specified directly from the SAM which is the original matrix of inter-industry transactions of all endogenous accounts. The form of matrix S corresponds to the form of matrix Z. S has the form

$$\begin{matrix} A & 0 & C \\ V & 0 & 0 \\ 0 & Y & H \end{matrix}$$

where A is the matrix of inter-industry technical coefficients, C is the matrix of endogenous final expenditure coefficients, V is the matrix of endogenous value-added input shares, Y is the matrix of endogenous coefficients distributing income to value-added categories, and H is the matrix of endogenous coefficients for distributing institution and household income. Equation (5.12) only gives the overall effect of change in exogenous demand on total output. To evaluate the direct, indirect, and feedback impacts of any exogenous injection, a method of decomposing the multiplier matrix M_A is used. The accumulated effects of S can be decomposed into three coefficient matrices, M_1, M_2, and M_3. So the equation for calculating the impact of exogenous injection on total output is now written as

$$x_i = M_A f = M_1 M_2 M_3 \times f \tag{5.13}$$

where M_1 is an $(n \times n)$ square matrix of direct-effect multipliers similar to Leontief output multipliers, M_2 is an $(n \times n)$ square matrix of indirect-effect multipliers which gives the effects of exogenous inputs of each type transmitted to the households, M_3 is an $(n \times n)$ square matrix of cross or closed-loop multipliers which captures the feedback effects of changes in household incomes on commodity consumption. The matrices M_1, M_2, and M_3 are calculated as follows:

We begin with the matrix $S = \begin{bmatrix} A & 0 & C \\ V & 0 & 0 \\ 0 & Y & H \end{bmatrix}$ discussed previously.

The matrix S is disaggregated into two additive matrices Q and R so that $S = Q + R$ where

$$Q = \begin{matrix} A & 0 & 0 \\ 0 & 0 & 0 \\ 0 & 0 & H \end{matrix} \text{ and } R = \begin{matrix} 0 & 0 & C \\ V & 0 & 0 \\ 0 & Y & 0 \end{matrix}$$

The vector of total output x is given by $\bar{x} = \begin{bmatrix} x \\ v \\ y \end{bmatrix}$ and the exogenous demand

vector is given by $\bar{x} = \begin{bmatrix} f \\ w \\ h \end{bmatrix}$ where x is the vector of total outputs, v is the vector of total value added, and y is the vector of total household income; f is the vector of exogenous final demand, w is the vector of exogenous value-added income, and h is the vector of exogenous household income. Equation 5.13 can be rewritten as

$$\begin{aligned} \bar{x} &= (R + Q)\bar{x} + \bar{f} \\ \bar{x} &= (I - Q)^{-1} R\bar{x} + (I - Q)^{-1}\bar{f} \end{aligned} \tag{5.14}$$

If $T = (I - Q)^{-1} R$, Eq. (5.14) becomes

$$\begin{aligned} \bar{x} &= T\bar{x} + (I - Q)^{-1}\bar{f} \quad \text{or} \\ T\bar{x} &= \bar{x} - (I - Q)^{-1}\bar{f} \end{aligned} \tag{5.15}$$

Multiplying Eq. (5.15) throughout by T, we get

$$\bar{x} = T^2\bar{x} + T(I - Q)^{-1}\bar{f} + (I - Q)^{-1}\bar{f} \tag{5.16}$$

The equation is expanded further by multiplying once again by T and substituting the result back into Eq. (5.14):

$$\begin{aligned} \bar{x} &= T^3\bar{x} + T^2(I - Q)^{-1}\bar{f} + T(I - Q)^{-1}\bar{f} + (I - Q)^{-1}\bar{f} \\ \bar{x} &= (I - T^3)^{-1}(I + T + T^2)(I - Q)^{-1}\bar{f} \end{aligned} \tag{5.17}$$

$$\begin{aligned} M_1 &= (I - Q)^{-1}\bar{f} \\ M_2 &(I + T + T^2) \quad \text{and} \\ M_3 &= (I - T^3)^{-1} \end{aligned}$$

A detailed discussion of decomposed SAM multipliers can be found in Miller and Blair (2009). The total production in each sector i in the economy can be now calculated using Eq. (5.13) $\rightarrow x_i = M_A f = M_1 M_2 M_3 \times f$

Structural path analysis (SPA) is a method through which the whole network through which the influence of an injection into the system is transmitted (as represented by the social accounting matrix framework) can be identified and analyzed. The traditional multiplier decomposition method provides three multiplier matrices: M_1—direct-effect multipliers, M_2—indirect-effect multipliers, and M_3—feedback-effect multipliers. These matrices in a multiplicative relationship provide the means to calculate the total production required in a sector based on change in the exogenous final demand given in Eq. (5.13) discussed in the previous section—$x_s = M_1 M_2 M_3 \times f_s$.

Here, M_1 is the direct-effect multiplier matrix. It includes Leontief output multipliers, but does not include multiplier effects associated with other sectors such as value added or households. M_2 is the indirect-effect multiplier matrix which includes indirect multipliers and records how the effects of exogenous inputs of each type get transmitted to the household sector. M_3 is the feedback-effect multiplier matrix which captures the feedback effects between households and inter-industry transactions.

The change in production can be similarly calculated using Eq. (5.18):

$$\Delta x_s = M_1 M_2 M_3 \times \Delta f_s \tag{5.18}$$

This can alternatively be written as an additive relationship as shown in Eq. (5.19):

$$\Delta x_s = \left[I + M'_1 + M'_2 + M'_3 \right] \times \Delta f_s \tag{5.19}$$

where $I =$ Actual injection into the sector:

$$M'_1 = M_1 - I$$
$$M'_2 = (M_2 - I)M_1$$
$$M'_3 = (M_3 - I)M_2 M_1$$

This provides a method to separate the effects of an exogenous injection into the economy into direct, indirect, and feedback effects.

5.3 Methodology Used for Constructing a Model Using the SAM

The modeling framework used in this analysis can be explained by dividing it into two sub-components. The first sub-component consists of the analysis done using structural path analysis referred to here as input-output analysis. It includes the social accounting matrix, the inputs to the SAM, the assumptions and the interlinkages between values from the SAM and values from outside SAM, and the calculations done to obtain values for the target year. The second sub-component consists of the conversion of values from monetary terms obtained from the input-output analysis to physical terms. This analysis is done for the electricity sector.

The SAM multipliers are calculated from the endogenous accounts of the SAM. The factor accounts of the SAM provide the value added in each sector of the economy. To achieve a certain growth rate between the base and target year in each sector these values of the labor and capital account should change accordingly. However, since these accounts are endogenous, they cannot be directly changed. In optimization models, parameters are fixed from outside to ensure that certain overall growth rates are achieved. Sectoral growth rates are almost never specified in the results of these models. However, in the input-output model presented here, to achieve desired sectoral growth rates, the inputs (or injections) into the exogenous accounts have to be adjusted appropriately. This is done by trial and error. So for example, to achieve a certain growth rate in the industrial sector, new capital will need to be created in this sector. Therefore, there will be an injection into the capital account of the industry sector. This injection in turn will lead to a change in the value added in the service and agriculture sectors as well. This is the reason that trial and error is required in order to get the desired rates of growth in value added in all sectors of the economy. The change in vector (f), which is the exogenous final demand vector, leads to a change in the total production required for that particular year, calculated using Eq. (5.13). The methodology has been illustrated in Fig. 5.2.

A perturbation in the capital account or the government expenditure account, i.e., a perturbation in the vector f, leads to a change in total value of goods and services produced in the economy, i.e., x. Two sectors, government expenditure and capital formation, are considered to be exogenous in this model as mentioned before. The values in these two sectors for each economic activity are adjusted in a specific ratio such that the total value added in each sector is arrived at. This is done manually by trial and error. In order to limit the arbitrariness introduced because of this trial and error, it is assumed that the ratio of government expenditure to total capital formation will remain constant. From the values of x obtained through this analysis a new SAM for the target year is estimated. The values of x are calculated using the matrices M_1, M_2, and M_3.

The values from the endogenous accounts in the SAM—i.e., the production activities, factors of production, households, and rest of the world accounts—make up the matrix S. This matrix can be further subdivided and matrices M_1, M_2, and M_3 can be estimated. Over a modeling time period which is more than say 3–4 years, the ratios (or values in the matrices M_1, M_2, and M_3) will change. The values of M_1 represent the direct effects of an injection into an endogenous account. The values for all other accounts change appropriately based on ratios in M_1. Now, this holds true even for the energy sectors. The amount of energy (in monetary terms) needed in the production of a unit of output in all other production sectors is obtained from the energy account rows of matrix M_1. In the event of an improvement in energy efficiency, these values in matrix M_1 will reduce. The change in the energy ratios of matrix M_1 follows Eq. (5.20):

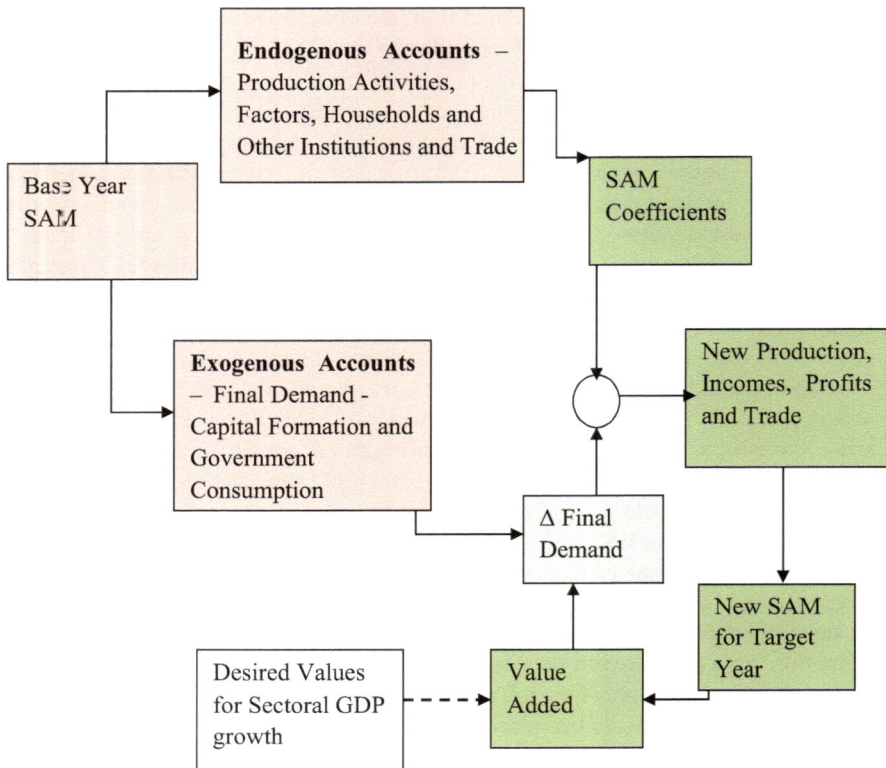

Fig. 5.2 Calculating total sectoral output and value added in each sector using a SAM framework

$$M_{1i\text{modified}} = (1 - \varepsilon_i) \times M_{1i\text{original}} \tag{5.20}$$

where i = energy sector rows, ε_i = annual energy efficiency improvement in energy sector i, and N = the total time period for the analysis.

Similarly we have to also account for change in factor productivity. For example, between 2003–2004 and 2030, the amount of labor and capital required to produce one unit output of say cement will have changed. Labor and capital use both will have become more productive. The effects of changes in factor productivity are captured in matrix M_2 and are shown in Eq. (5.21):

$$M_{2i\text{modified}} = M_{2i\text{original}} \times (1 - \rho_i) \tag{5.21}$$

where i represents the factor rows, and ρ_i is the annual improvement in factor productivity.

The matrix M is then calculated using Eq. (5.22):

$$M = M_3 \times M_2 \times M_1 \tag{5.22}$$

One of the main criticisms of the Leontief input-output analysis is that these multipliers are taken as the basis on which all forecasting is done, without accounting for the fact that these do not allow for substitution and scale effects. In this analysis this is addressed by allowing the multipliers to vary within a specified range. This range is obtained from actual data; that is, the structural linkages between the same two sectors in 1993–1994, 2003–2004, and 2007–2008 at constant prices are calculated and the lowest and highest values provide the range within which the value can vary. This may not completely address the limitations of the method, but it provides a way in which some flexibility can be introduced in the model without adding the complexity and uncertainty of an optimization, as in the CGE models.

The GDP for the base year is calculated from the SAM. The SAM provides the values for annual "value added" in each productive sector by both labor and capital. The total value added in year t can be calculated from the SAM using Eq. (5.23):

$$VA_t = \sum_{i=1}^{n} \left(VA_{i,t,\text{Labour}} + VA_{i,t,\text{Capital}} \right) \tag{5.23}$$

where VA_i = the value added in sector i, and n = number of sectors in the economy.
The growth rate of each sector is then calculated using Eq. (5.24):

$$GR_i = \exp\left\{ \left[\ln \frac{VA_t}{VA_{\text{base}}} \right] / [t - t_{\text{base}}] \right\} - 1 \tag{5.24}$$

where GR_i is the growth rate in sector i.
The value added for the target for each sector is obtained from the SAM from the target year. The SAM for the target year is constructed using the new values of production x given by vector $[x]$, new values of exogenous final demand f given by vector $[f]$, and the matrix S which is available from the base year SAM.
The following equations are used:

$$SAM_{\text{new}} = Z_{\text{new}} + f_{\text{new}} \tag{5.25}$$

$$Z_{\text{inew}} = S \times \hat{x}_{\text{new}}^{-1} \tag{5.26}$$

where Z is the matrix of inter-industry and other endogenous transactions and $\hat{x}_{\text{new}}^{-1}$ is the square matrix representing the total output vector. The vector f is the vector of exogenous final demands. This vector is modified in order to get the desired growth rates in each sector. "f_{new}" is the new exogenous final demand. The sectors that have been specified as exogenous in this analysis are those of government consumption and capital formation. Therefore f_{new} is given by Eq. (5.27):

$$[f_{new}] = [GE^{new}] + [CF^{new}] \tag{5.27}$$

where GE^{new} is the new government demand/expenditure in the target year and CF^{new} is the capital formation in the target year. The exogenous final demands can be calculated using Eq. (5.28):

$$F_i^{new} = \frac{\left(F_i^{old} + a_i\right) \times n}{(1 + D)^n} \tag{5.28}$$

where "a_i" is the factor by which the exogenous final demands in sector i change, "D" is the discount rate, and "n" is the total number of years. Usually economic analysis based on input-output matrices may stop here; however for our purposes we also need to convert the results obtained in physical terms from the input-output analysis into physical units. The second sub-component of the input-output model is the conversion of monetary values to physical terms for a few key sectors—i.e., energy, cement, and iron and steel. For the cement and iron and steel sectors, no separate analysis is done to evaluate different technologies used, etc.; the conversion to physical terms is done so as to get some idea of the physical requirement of the commodity.

For the power sector however, this conversion to physical terms is done so that it can be connected to a separate technology-explicit analysis of the power sector. So, for a scenario constructed in the input-output model by perturbing the values for the exogenous sectors, i.e., government expenditure and capital formation, the output will be the total production of commodities and services for a target year. Of these commodities, electricity is one. This electricity demand is then converted to physical terms. This is done by using a conversion factor which is essentially the price of electricity, i.e., $/kWh. For the base year, the data for actual electricity consumed in physical units by each sector is available from the CEA reports and the amount in value terms is available from the SAM. The ratio of the two values is the conversion factor for the base year. Now it can be assumed that this remains the same or changes between the target and base years. For the purpose of this calculation the rate of change in this ratio between the years 2003–2004 and 2007–2008 is assumed to be the rate at which this ratio changes between the base and target years. Using this method, based on the new SAM constructed for the target year from the new values of production, a value of energy demand in physical terms is estimated. Now for this energy demand, a fuel mix has to be determined. This is done by the power sector model which estimates the optimal fuel mix to meet a given demand under multiple constraints. The power sector model has been discussed in the previous chapter.

For any particular fuel mix, a certain amount of cumulative and annual investment will be needed. This investment is fed back into the exogenous capital account of the SAM as $\Delta CF_{i = power}$, i.e., capital required in the power sector. In the absence of any constraints on the power sector—i.e., constraints on meeting renewable energy targets or constraints on emissions, the decision about the optimal fuel mix to supply the energy demand is made based solely on minimizing the energy and capital costs. In this case the structural linkages between various sectors in the economy represented by the SAM coefficients determine the distribution of government expenditure and no exogenous/manual adjustment is made to the government expenditure head in the SAM. In this case (let us call this scenario "A"), the total exogenous injection in the economy is given by Eq. (5.29):

$$\sum_i \Delta GE_i^A + \sum_i \Delta CF_i^A = I^A \qquad (5.29)$$

where I^A is the total exogenous injection into the economy in scenario "A." Additional constraints such as one on emissions, or one on the amount of renewable energy mandated to be in the system, would require a different fuel mix and would therefore lead to a different result for total investment required. This investment required in the power sector is represented by the total new capital formation (in terms of value) in the power sector between the base and target year, in the SAM. Therefore in the new scenario (say "B") in which additional investments would be required in the power sector because of constraints on emissions, say, the total exogenous injection in the economy would change as given by Eq. (5.30):

$$\sum_i \Delta GE_i^A + \sum_i \Delta CF_i^A + \Delta CF'_{i=power} = I^B = I^A + \Delta CF'_{i=power} \qquad (5.30)$$

where I^B is the new exogenous injection required in the economy under scenario "B" which is increased by a factor equivalent to $\Delta CF'_{i=power}$ that is the new capital required in the power sector under constraints. In the absence of any exogenous adjustment, this additional investment will be an extra injection into the economy. This means that if in scenario "A" the total exogenous injection into the economy is I, then in scenario "B" it will be $I + \Delta I$ where this ΔI will be equal to the total new capital formation required in the power sector ($\Delta I = \Delta CF'_{i=power}$). However, for the economy as a whole, and especially a developing economy which has to face severe capital constraints, the total available investment at any given point of time may not vary very significantly. Even if it is assumed that part of the new investment required in the power sector comes in the form of private investment and is injected into the economy over and above the other planned government and private expenditures, it is still reasonable to assume that a part of this investment may have to come from the government. And therefore, given that there is constrained budget of the government, a higher expenditure in one sector, in this case the power sector, may result in a reduction in expenditures in other sectors. This is shown by Eq. (5.31):

$$I^B = \sum_i \Delta GE_i^A + \sum_i \Delta CF_i^A + \Phi_g \Delta I + (1 - \Phi_g)\Delta I \qquad (5.31)$$

where ΔI = total new exogenous injection into the economy, which is equal to the new capital required in the power sector, $\Delta I = \Delta CF'_{i=power}$. Φ_g is the fraction of total investment that comes from the government budget. The investment that is made by the government which presumably works within a specified budget then has to be sourced from other expenditure heads. Therefore this additional term of $\Phi_g \Delta I$, which is the additional investment that the government makes in the power sector has to be deducted from other sectors of the economy if we assume that the total expenditure that the government can undertake is constrained. Equation (5.32) is therefore modified to give $I^{B'}$ which is in effect $I^B - \Phi_g \Delta I$. Equation (5.32) represents the operation form of this calculation:

$$I^{B'} = I^B - \Phi_g \Delta I = \sum_i (GE_i^A - \Phi_{ig} \Delta I) + \sum_i CF_i^A + (1 - \Phi_g) \Delta I \qquad (5.32)$$

where Φ_{ig} is the fraction of government expenditure on the power sector that is deducted/sourced from sector i. Here $\sum_i \Phi_{ig} = \Phi_g$. The next key variable therefore is the source of this additional government investment—i.e., whether the government reduces welfare expenditure, that is, say, expenditure from sectors such as health and education which in this analysis are clubbed under the "service" sector, or expenditure in other sectors. As shown in Fig. 5.1 there is a feedback loop between the power sector module and the SAM module which enables the incorporation of different investments corresponding to different fuel mixes into the SAM.

5.4   SAM 2003–2004 and Other Data

Two SAMs are available in the open domain. One is a SAM for the year 2003–2004, constructed by Saluja and Yadav (2006) and published by the Planning Commission (Saluja and Yadav 2006). Another more recent SAM for the year 2007–2008 is also available. It is constructed by a team at the Institute of Economic Growth as part of the "11th Five Year Plan and Beyond" program supported by the Planning Commission and "Inclusive Growth in India" project of the Norwegian Government. The 2007–2008 SAM is more recent and therefore using it may provide results that are more accurate. However, there were two disadvantages in using this SAM. The classification of household classes in the 2007–2008 SAM is based on occupations and not income (or expenditure) classes. This makes it difficult to actually evaluate any measure of equality as there is a wide income range within every occupational category. The results for income distribution are thus skewed—inequality is undercalculated. The other disadvantage is that the results obtained from using the SAM 2007–2008 cannot be validated against some other studies as most other studies for energy and emission projections for the future have used the SAM 2003–2004. In this analysis the SAM for 2003–2004 constructed by Saluja and Yadav is used.

The methodology for the construction of the SAM 2003–2004 is discussed in detail in Saluja and Yadav (2006). The SAM has 73 production sectors and 73 commodities. The final demand sectors comprise households, government, capital formation, and rest of the world. The households are divided into ten household groups or classes—five rural and five urban. This classification is a slightly modified version of the expenditure-wise classification given in the NSSO survey for 1999–2000 (NSS Report 461). Two classes from the NSS survey are combined to form one household class in the SAM. For our purpose the SAM has been aggregated to nine production sector and nine commodities. These are agriculture, cement, iron and steel, other industry, services, coal and lignite, crude oil and natural gas, petroleum products, and electricity. As is evident, the energy sectors—both primary and secondary—are separate and have not been aggregated in the other industry category. The cement and iron and steel sectors are separated out as it is possible to get physical measures of these commodities. So the monetary analysis using the SAM can be connected to physical measures of production.

The classification of household classes in SAM 2003–2004 provides insights into the incomes, income shares of the different household groups, and their expenditures on different types of commodities. It provides a measure of the inequality among households as well as insights into the sources of income for each household class. The data shows that the income shares of the poorest household groups (both rural and urban) are significantly lower relative to their share of the total population as opposed to the richest rural and urban groups whose income shares exceed their population shares by a high margin. Using this data it is possible to measure inequality among the household classes using either the Gini or the Theil index.

The GINI coefficient is a simple measure of inequality between groups but is less suited for measuring inequality within groups, mainly because it is not additively decomposable into between- and within-group inequality (Bourguignon 1979). The GINI is calculated assuming perfect equality of income within household groups for the household classes given in SAM 2003–2004. The GINI coefficient is calculated using Eq. (5.33):

$$\text{GINI} = \sum_{h=1}^{N-1} \eta_{h+1} \text{FP}_h - \sum_{h=1}^{N-1} \eta_h \text{FP}_{h+1} \tag{5.33}$$

where η_h is the cumulative income share of household group h and FP_h is the cumulative population share of household group h. The Theil index is more useful if a further disaggregation is available to measure intra-class inequality. For example, if data is available within each household class in terms of, say, occupational segregation (e.g., farm labor vs. non-farm manual labor) or caste-based segregation or segregation based on levels of education, further insights into the inequality among households can be obtained using the Theil index. Some work in this direction has been undertaken by Pieters (2010). Household classes in the analysis done by Pieters (2010) have been segregated based on three levels of education. However the constraint on obtaining data makes it very difficult to undertake such an exercise. For our purpose we have only calculated the GINI index as further segregation of income classes was not available. For the year 2003–2004, from the data obtained from this SAM, the GINI coefficient is calculated as 0.4973.

The income of each household class is spent on the purchase of commodities and on paying taxes. A part of the income is saved. Income received by households also varies in terms of the sources of income. The income can be a direct endowment for services rendered in national production activities, government transfers, or income earned and transferred from abroad. The poorest household classes (UH1, RH1, UH2 and RH2) do not earn any income from abroad. Of the total government transfers to households, about 10% is transferred to these four household groups. The high-income household classes (UH5 and RH5) are the beneficiaries of 49% of total government transfers to households in the form of interest returns and subsidies.

Of the total income earned by households, a certain percentage is spent on final consumption of commodities and services, some percentage is paid to the government as direct or indirect taxes, and the rest is saved. Almost half of the income of the household group RH5 is kept as savings. In comparison, the corresponding urban household group (UH5) saves a lesser percentage of its income and spends a higher percentage of its income on services. The household groups UH1 and RH1 have net negative savings; that is, these household groups are in debt. The poorest rural household group (RH1) spends about 61% of its income on agricultural products and about 27% of its income on services whereas the richest urban household group (UH5) spends about 16% of its income on agricultural products and 71% of its income on services. All the urban household groups have to spend a significant amount of their income on housing. For the poorest urban household group, of its total expenditure, 15% is spent on housing. The high expenditure on services by groups UH3, UH4, UH5, and RH5 is mostly accounted for by expenditure on health and education services.

5.5 Illustration of the I-O Analysis Through Scenarios

The SAM in this analysis has not been used to produce predictions for energy requirements or emissions for the future. Although other models that use the SAM claim to do this, we believe that to claim such an outcome is erroneous as the SAM results are based on many assumptions made at almost every level of analysis. However a SAM-based model is useful in evaluating "what if"—scenarios built to evaluate impacts of certain policy decisions on key developmental parameters. For example, what is the impact of high penetration of renewable energy technologies on income equity and employment? To evaluate such impacts on key development indicators, we have built three scenarios. The time period for the study is taken to be 2004–2030. Many studies commissioned by the Government of India use 2030 as the benchmark target year (IEP, Low Carbon Report). The INDC (Intended Nationally Determined Contributions) submitted by the Government of India also sets targets for the year 2030. Therefore using this target year provides a basis for comparison with government targets.

The first scenario (called the business-as-usual or BAU scenario) is a baseline scenario. This does not necessarily represent what will actually happen in the future, but with many parameters remaining unchanged relative to the base year this scenario can be used as a benchmark against which other scenarios can be evaluated. In the BAU scenario it is assumed that the sectors, viz., agriculture, industry, and services, grow at the same average rates of growth as witnessed in the decade 1998–2008. For agriculture this means a growth rate of 2.7%, a growth rate of 6.6% in industry, and a growth rate of 8.1% in the service sector. The production of steel is estimated to reach 140 million tonnes and the production of cement to reach 420 million tonnes by 2030. For steel and cement, these estimates are taken from planned increase in production estimated by the respective ministries. In the electricity sector, it is assumed that the total contribution of each fuel technology to electricity generation is as at the level it has achieved in 2014. This means that coal dominates electricity production followed by natural gas. Renewable energy, not including large hydropower plants, plays a smaller role in total electricity generation. This baseline thus constructed is then kept constant with changes in only some parameters to evaluate the impact of policy decisions on key parameters of development.

The next scenario (Scenario I) is a renewable energy scenario in addition to the BAU scenario. There are no emission restrictions in this scenario. However, it is assumed that the contribution of renewable energy sources to total electricity generation will increase significantly to about 15% of total electricity generation (not including electricity generated by large hydropower plants). It is also assumed that about 63 GW of nuclear power will also be available by 2030. This entire increase in renewable energy generation is assumed to replace coal-based generation. No other parameter is altered. The impact of a change in fuel mix on GDP growth, incomes, equality, and employment can be evaluated vis-à-vis the BAU scenario.

Scenario II is a poverty alleviation scenario. In this scenario, all assumptions of Scenario I are kept constant and it is assumed that for households that are below the poverty line of 2005 constant 2\$/person/day at PPP, there are direct government transfers in order to make sure that their income increases to above the poverty line threshold. The impact of such transfers on GDP, equality, energy requirements, etc. is then evaluated. The key parameters used to construct the scenarios are described in Table 5.2.

The results obtained from running the SAM model for these scenarios are discussed in the next section.

5.6 Results and Discussion

With the sectoral growth rates at the values discussed above, an overall growth rate of 6.94% is achieved in the BAU scenario. Some key results of the BAU scenario are given in Table 5.3 in comparison to values of the same parameters in base year 2003–2004.

Table 5.2 Scenarios built for analyzing impacts of renewable energy deployment on economic indicators

Scenario	Sectoral growth rates (annual %)			Physical units—production			Electricity production		Government transfers
	Agriculture	Industry	Services	Iron and steel (Mt)	Cement (Mt)		Contribution of fossil fuels (%)	Contribution of renewable energy technologies (%)	Y/N
BAU	2.7%	6.6%	8.1%	140	420		83%	4%	N
Scenario I	Output	Output	Output	Output	Output		67%	15%	N
Scenario II	Output	Output	Output	Output	Output		67%	15%	Y

Table 5.3 Comparison of base year and BAU scenario for target year 2030

	Base year—2003–2004	BAU—2030
GDP (Rs. trillion)	26.5	151.8
GDP growth p.a.	–	6.94%
Per capita income of all HHds (Rs. 1000/year/person)	21.6	85.2
GINI coefficient	0.4974	0.5291
% of population spending less than $2/person/day (@Rs.20/$—PPP)	59%	12%

Table 5.4 Energy demand, emissions, and investment in the power sector in BAU scenario

	Base year—2003–2004	BAU—2030
Per capita electricity demand (kWh/person/year)	634	1974
Total installed capacity (GW)	168	619
Annual investment in electricity sector (Rs. billion)		964
Per capita emissions (tCO$_2$/person/year)	1.66	5.07

In the BAU scenario, the GINI coefficient increases; that is, inequality and unemployment increase even as per capita income increases. However population below the poverty line of $2/person/day reduces though 12% of the population is still under the poverty line. These values are only indicative in that they represent the baseline trend between the base year and target year that is used in this analysis in relation to which other scenarios are compared. The energy variables for this scenario are shown in Table 5.4.

The total emission flow in year 2030, in the BAU scenario, is expected to reach 7.5 GtCO$_2$ from a value of 1.9 GtCO$_2$ in the base year. The United States currently (2012) emits about 5.12 GtCO$_2$ (excluding land-use change and forestry) (WRI-CAIT Database, 2014). As compared to developed countries and even mid-level developed countries, the energy consumption of 1974 kWh/person/year is less. However even this requires an annual investment of Rs.964 billion in the power sector. The working group on power for the 12th 5 Year Plan had projected an investment of about Rs. 13,72,580 crores during the 12th plan period of which about 12,35,000 crores is earmarked for generation. This amounts to about Rs. 2471 billion per year. This amount is substantially high as compared to the investment required as projected by our model (Rs. 964 billion). It should be remembered however that the planning commission figure is only a projection of requirement and it does not reflect the amount that will actually be invested in this sector. The latter is more likely to be closer to the Rs. 964 billion projected by the model. In the 11th 5 Year Plan the investments in the power sector were to the tune of about Rs. 1200 billion (inclusive of generation, transmissions, and distribution).

In Scenario I all other parameters are left unchanged as compared to BAU and only the mix of technologies used in the power sector is changed. Instead of a very low contribution by renewable energy technologies as assumed in the BAU scenario, it is assumed that electricity generation from renewable energy sources (solar, wind, biomass and small hydro power) would reach 15% of total electricity generation and that from all nonfossil energy sources (including large hydro and nuclear) would reach 33% (one-third of total electricity generation by 2030). An additional investment will be required for this. This investment can come from either private developers or the government. Private investment of the scale required for this transition will be possible however only if the government provides a conducive policy environment which may include both capital and operational subsidies indirectly affecting government expenditure. In case of a fully government-funded program, in a budget-constrained economy, this additional investment will have to be diverted from other accounts. It will depend on the policies of the incumbent government as to how this is done. In this scenario, it is assumed that 50% of the new investment is made by the government and the remaining 50% comes from the private sector. It is also assumed here that the extra government investment required in the power sector is met by reducing investments in all other sectors of the economy, in the proportion of their contribution to the GDP. For example, say the total additional capital formation in the power sector to meet energy demands is Rs.100 billion. This extra Rs.100 billion requirement will have to be met from other sectors. So the iron and steel sector which contributes 1.5% to the total GDP of the country will get Rs. 100 billion × 1.5% = Rs. 1.5 billion less investment in the target year. In such a case the impacts on GDP growth, incomes, and equality are evaluated. It is of course an assumption that the extra investment is compensated by all other sectors. Different scenarios can be run to see what happens if say all of the extra investment is taken from one sector alone or in a higher proportion from some sectors as compared to others (this is demonstrated in the next chapter). A comparison of the two scenarios BAU and Scenario I described above in terms of the energy parameters in the two scenarios is given in Table 5.5.

A higher percentage of renewable energy technologies results in the reduction of per capita emissions by 13%. This reduction however is achieved at a higher cost—about 42% increase in the required annual investment. In a budget-constrained economy this increase is met by reduced investments in other sectors. Therefore while there is a positive effect on GDP of increased investments in the power sector, there is a negative effect due to decreased investments in other sectors. The impact of a higher penetration of renewable energy technologies on economic parameters is shown in Table 5.6.

Table 5.5 Comparison between BAU and Scenario I

	BAU— 2030	Scenario-I— 2030
Electricity generation by fossil fuels (% of total generation)	83%	67%
Electricity generation by renewable energy technologies (% of total generation)	4%	14%
Total installed capacity (GW)	619	744
Annual investment in the electricity sector (Rs. billion)	964	1374
Per capita emissions (tCO$_2$/person/year)	5.07	4.39

Table 5.6 Comparison between BAU and scenario I—economic parameters

	BAU—2030	Scenario I—2030
GDP (Rs. trillion)	151.8	151.2
Total income of all HHds (Rs. trillion)	125.8	124.1
GINI coefficient	0.5291	0.5339
% of population spending less than $2/person/day (@Rs.20/$—PPP)	11.5%	11.5%

In a budget-constrained economy, the introduction of renewable energy technologies has a negative impact on the GINI coefficient. This means that the inequality increases due to the introduction of a higher percentage of renewable energy technologies. As compared to the base year, there is an increase in income of all classes in both the scenarios. However, because of the differential sectoral growth rates and the differential way in which the household classes are dependent on the sectors, the growth in incomes is also differential. In the BAU scenario, for example, the per capita income of household class UH1, which is the poorest urban household class, increases by 36% with respect to the base year whereas in the same scenario the income of household class UH5 which is the richest urban household class increases by 301%. On the other hand, in Scenario I, the income of the household class UH1 increases by only 4% whereas the income of household class UH5 increases by 299% with respect to the base year. While there is a decrease in the incomes of both household classes in Scenario I as compared to BAU, income of class UH1 has decreased by 23% while the income of household class UH5 has only decreased by 1%. A closer evaluation of the difference between the two scenarios can be done using structural path analysis (SPA). The effects can be subdivided further into direct, indirect, and feedback effects and the total activity impact can be disaggregated by sector in order to understand its independent components. The actual exogenous injection in the economy, in the BAU scenario, comes to about Rs. 49 billion. This injection in turn causes a direct increase in total economic activity amounting to about Rs. 127 billion. The indirect effect on economic activity amounts to Rs. 133 billion, and the increase due to feedback effects is Rs. 457 billion. The total activity impact of this injection is therefore Rs. 766 billion.

In Scenario I the exogenous injection remains the same as it represents the total budget and non-budget expenditure of the state as well as anticipated investments by the private sector within some constraints. However, the structure through which this injection now passes under Scenario I is different as renewable energy deployment in the power sector would require a larger investment in this sector leading to lower investments in other sectors. The total direct, indirect, and feedback impacts on economic activity amount to Rs. 133 billion, Rs. 130 billion, and Rs. 455 billion, respectively.

In Scenario II, it is assumed that the government has a policy objective of complete poverty alleviation by 2030—which means that by 2030, no one is living on less than $2/person/day. This is ensured by direct government transfers to

Table 5.7 Comparison between BAU, Scenario I, and Scenario II

	BAU— 2030	Scenario I— 2030	Scenario II— 2030
Total income of all HHds (Rs. trillion)	126	124	128
GINI coefficient	0.5291	0.5339	0.5210
% of population spending less than $2/person/day (@Rs.20/$—PPP)	12%	12%	0%
Per capita electricity demand (kWh/person/year)	1974	1945	1985
Total installed capacity (GW)	619	744	759
Annual investment in the electricity sector (Rs. billion)	964	1374	1410
Per capita emissions (tCO$_2$/person/year)	5.07	4.39	4.48

households that are under this poverty threshold. In such a case the changes in key energy and economic parameters are captured and shown in Table 5.7. All other assumptions (other than the government transfers to households below the poverty line) are kept identical to the assumptions made in Scenario I. Table 5.7 therefore shows the comparison between BAU, Scenario I, and Scenario II.

The GINI coefficient reduces as transfers to lower income households lead to an increase in per capita incomes. Between Scenario I and Scenario II, the increase in extra investment needed in the power sector to ensure both high renewable energy penetration as well as full poverty alleviation is not very substantial. The analysis indicates that a high renewable energy deployment scenario does not implicitly imply development and growth as argued by some studies (Pollin and Chakraborty, 2015). Depending on the manner in which the investment for renewable energy deployment is obtained, there could be a negative impact on the economy as well. The objective of poverty alleviation and development in the well-being of the population should therefore be considered as an independent policy priority and not be implicitly linked to achieving "green growth."

5.7   Conclusions

The model discussed in this chapter uses structural path analysis to analyze the impacts of a particular change in investments and expenditures on total economic activity. Input-output models are an established method of evaluating economy-wide impacts of policy decisions. There are studies for example that have used the standard Leontief I-O/SAM analysis to calculate emissions for 1 year for the Indian economy, the most notable being the paper by Parikh et al. (2009). These provide an overall account of the direct and indirect energy consumption and emissions of the Indian economy. The model developed here extends this analysis and connects the input-output analysis to a detailed analysis of the electricity sector. This enables an evaluation of different technologies and the technical and financial constraints in the power sector. This analysis also serves as an illustration for undertaking similar analysis for other end-use sectors. For example, the transport sector can be similarly analyzed and connected with an input-output model to enable a user to construct multiple scenarios and understand better the role of a sector in the economy as a whole.

The model developed here enables the user to construct multiple scenarios for the power sector and quantitatively estimate the impact of any decision in the sector on other sectors of the economy, on economic growth, and on household incomes. The results for three scenarios that were built to demonstrate the model show that a higher share of renewable energy in the total fuel mix, as in Scenario I, leads to a welfare loss in terms of consumption and a GDP loss as compared to the business-as-usual scenario. This is contingent on the set of assumptions about the structural linkages in the economy, change in fuel prices, and the way in which the additional investment required is obtained. There are various ways in which the deployment of large amounts of renewable energy can affect the economy. If most of the new investment in renewable energy technology comes from private institutions, then the effect on economic growth would be different than if most of the investment comes from the government. For each of these scenarios, there is a possibility of balancing the impact of an investment or divestment in the economy. This can be identified using structural path analysis as it enables us to trace the path through which the effect of an investment passes through the economy.

While there is very often an implicit assumption that there is a negative trade-off between the environment and development, the nature of this trade-off as well as its quantum is seldom estimated. The model presented here provides a methodology to evaluate and quantify this trade-off. The next chapter discusses the integrated modeling framework in which the input-output model is linked to the power sector model and the decomposition analysis in one framework.

References

Apergis, N., & Payne, J. E. (2014). Renewable energy, output, CO_2 emissions, and fossil fuel prices in Central America: Evidence from a nonlinear panel smooth transition vector error correction model. *Energy Economics, 42*, 226–232.

Bourguignon, F. (1979). Decomposable income inequality measures. *Econometrica, 47*, 901–920.

CAIT Climate Data Explorer. (2014). Washington, DC: World Resources Institute. Retrieved from https://cait.wri.org.

Del Rio, P., & Burguillo, M. (2009). An empirical analysis of the impact of renewable energy deployment on local sustainability. *Renewable and Sustainable Energy Reviews, 13*(6–7), 1314–1325.

Miller, R. E., & Blair, P. D. (2009). Input-output analysis: foundations and extensions. Cambridge university press.

Myles, H. (1998). *The socio economic impact of renewable energy technologies*. Guildford: Surrey Energy Economics Centre.

Parikh, J., & Ghosh, P. P. (2009). Energy technology alternatives for India till 2030. *International Journal of Energy Sector Management, 3*(3), 233–250.

Parikh, J., Panda, M., Ganesh-Kumar, A., & Singh, V. (2009). CO_2 emissions structure of Indian economy. *Energy, 34*(8), 1024–1031.

Pieters, J. (2010). Growth and inequality in India: Analysis of an extended social accounting matrix. *World Development, 38*(3), 270–281.

Pollin R., & Chakraborty, S. (2015). An egalitarian green growth programme for India. *Economic & Political Weekly, 1*(42), 38–52.

Pyatt, G. (1988). A SAM approach to modeling. *Journal of Policy Modeling, 10*(3), 327–352.

Pyatt, G., & Round, J. I. (1985). *Social accounting matrices: A basis for planning*. Washington: The World Bank.

Reinert, K. A., & Roland-Holst, D. W. (1997). Social accounting matrices. In *Applied methods for trade policy analysis: A handbook* (pp. 94–121). Cambridge: Cambridge University Press.

Round, J. (2003). Social accounting matrices and SAM-based multiplier analysis. In *The impact of economic policies on poverty and income distribution: Evaluation techniques and tools* (pp. 261–276). Basingstoke: Palgrave Macmillan.

Saluja, M. R., & Yadav, B. (2006). *Social accounting matrix for India 2003–04*. Haryana: India Development Foundation.

Stern, N. H. (2006). *Stern review: The economics of climate change* (Vol. 30). London: HM treasury.

Chapter 6
An Integrated Modeling Framework (IMF) for Energy-Economy-Environment Modeling

Abstract Addressing environmental concerns alongside economic development and energy transitions is a challenge for all countries and effective policy making requires approaches that can balance all these concerns. In this chapter, combining a range of modeling methodologies is proposed as a way of addressing these questions in an integrated manner. The choice of models to be used would depend on the specific characteristics of the system and region being studied. One possible combination of models is discussed in this chapter, taking off from the modeling methods discussed in previous chapters, for the specific case of India. The integrated modeling framework (IMF) proposed combines three modeling approaches—(i) index decomposition to estimate impacts of structural changes in the economy, (ii) constrained optimization to estimate least-cost fuel options for the power sector, (iii) input-output analysis to estimate economic impacts. The model results indicate that the mode of investment for climate change mitigation is a significant determinant of the impact on economic growth, incomes, and income distribution in India. In some scenarios, higher investments in green energy negatively affect low-income households significantly more as compared to other households. The chapter also discusses how the IMF can be used to determine the reasons for and therefore alleviate the negative impacts.

Keywords Integrated modeling framework · Constrained optimization · Generation expansion planning · Input-output analysis · Equity and sustainability

The three models that are linked together in an integrated framework have been individually discussed in the previous chapters. In this chapter the integrated modeling framework—IMF—is discussed. A few scenarios are constructed and run using the IMF to illustrate the potential use of this framework.

© The Author(s), under exclusive licence to Springer Nature Switzerland AG 2020   89
T. Kanitkar, *An Integrated Framework for Energy-Economy-Emissions Modeling*,
SpringerBriefs in Environmental Science, https://doi.org/10.1007/978-3-030-18263-2_6

6.1   The Integrated Modeling Framework: IMF

The integrated modeling framework (IMF) proposed here combines the three models discussed in the previous chapters, viz., —(i) decomposition analysis, (ii) optimization, and (iii) input-output analysis. Figure 6.1 shows the schematic of the IMF.

Model 1 which is the decomposition analysis provides the analysis of past emissions which in turn provides the basis for building scenarios for the future. In Model 1, the emission intensity of GDP is decomposed into its component parts, shown in Eq. (6.1):

$$E = \sum_i I_i \times C_i \times S_i \times G \tag{6.1}$$

where I_i is the energy intensity of GDP for sector i, C_i is the emission intensity of energy for sector I, S_i is the contribution of sector i to total GDP, and G is the total GDP of the region. The analysis provides insights into the contribution of each factor to change in emission intensity of GDP in the past and to its potential role in the future. The potential ranges for two key variables can be obtained from this analysis—(i) the change in the structural composition of the economy, i.e., the range for value added in different sectors of the economy (ΔG_i), and (ii) the change in energy

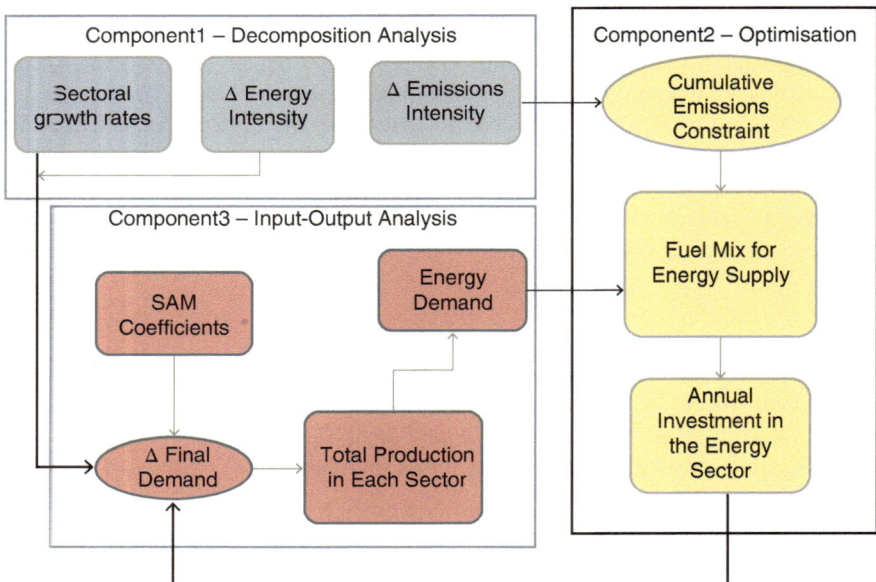

Fig. 6.1 Integrating three models—decomposition of past emissions, I-O analysis, and optimization; BY is base year and TY is target year

efficiency or use of energy in any sector represented by the energy intensity of GDP in each sector (ΔEI_i).

The results from Model 1 provide the limits for the values of these variables while constructing scenarios in Model 3. Model 3 is an extended I-O model calculated based on a social accounting matrix constructed for a base year. In this analysis through multiplier decomposition the total production of goods and services required in each sector in the target year is obtained for each scenario. Here, two final demand sectors—government expenditure and capital formation—are considered to be exogenous while all other sectors are endogenous to the model. The selection of the exogenous final demand sectors is based on convention and reflects the behavior and drivers of economic change in planned economies (Taylor and Taylor 2009). The values of consumption expenditure in these two components in each of the commodity sectors are adjusted till the change in value added in each sector matches the growth rates required for each scenario. Since this is done by trial and error, a set of values for GE and CF that result in a certain growth rate in each sector may not be unique. In order to reduce the arbitrary nature of these adjustments therefore, it is assumed here that the ratio of government expenditure to capital formation in each sector remains the same as in the base year. The new values for GE_i and CF_i are given by Eqs. (6.2) and (6.3):

$$GE_i^{TY} + GE_i^{BY} + \Delta GE_i \tag{6.2}$$

$$CF_i^{TY} + CF_i^{BY} + \Delta CF_i \tag{6.3}$$

The values for ΔGE_i and ΔCF_i are input by the user and represent the total change in government expenditure and capital formation, respectively. These values are linked through the SAM coefficients to all other sectors of the economy and therefore they can be adjusted such that the value added in each sector, i.e., the value of labor, capital, and indirect taxes in each sector, adds up to the total value added required for the scenario. The manner in which this is done is explained in detail in Chap. 5. However, a part of the discussion is reproduced here for reference. For a given set of growth rates in each sector, there is an energy demand in the target year that is generated as output from the input-output analysis. This energy demand takes into consideration the changes in energy efficiency in each sector. This energy demand which is the output of the input-output analysis is then given as input to the power sector model (Model 2 discussed in detail in Chap. 4 and summarized in a later paragraph).

One of the outputs from the power sector model is the investment required for a particular fuel mix that will supply the economy-wide energy demand. This investment is then fed back into the input-output model as a new $\Delta CF_{i \,=\, power}$, i.e., capital required in the power sector. In the absence of any constraints on the power sector, i.e., constraints on meeting renewable energy targets or constraints on emissions, the decision about the optimal fuel mix to supply the energy demand is made based solely on minimizing the energy and capital costs. In this case the structural linkages between various sectors in the economy represented by the SAM coefficients determine the distribution of government expenditure and no exogenous/manual adjustment is made to the government expenditure head in the SAM. In this case (let us call this scenario "A"), the total exogenous injection in the economy is given by Eq. (6.4):

$$\sum_i \Delta GE_i^A + \sum_i \Delta CF_i^A = I^A \qquad (6.4)$$

where I is the total exogenous injection into the economy in scenario "A."

Additional constraints such as one on emissions, or one on the amount of renewable energy mandated to be in the system, would require a different fuel mix and would therefore lead to a different result for total investment required. Therefore in the new scenario (say "B") in which additional investments would be required in the power sector because of, say, some constraints on emissions, the total exogenous injection in the economy would change as given by Eq. (6.5):

$$\sum_i \Delta GE_i^A + \sum_i \Delta CF_i^A + \Delta CF_{i\in power} = I^B = I^A + \Delta CF_{i\in power} \qquad (6.5)$$

where I^B is the new exogenous injection required in the economy under scenario "B" which is increased by a factor equivalent to $\Delta CF'_{i\,=\,power}$ that is the new capital required in the power sector under constraints. In the absence of any exogenous adjustment, this additional investment will be an extra injection into the economy. This means that if in scenario "A" the total exogenous injection into the economy is I, then in scenario "B" it will be $I + \Delta I$ where this ΔI will be equal to the total new capital formation required in the power sector ($\Delta I = \Delta CF'_{i\,=\,power}$). However, for the economy as a whole, and especially a developing economy, the total available investment at any given point of time may be constrained. Even if it is assumed that part of the new investment required in the power sector comes in the form of private investment and is injected into the economy over and above the other planned government and private expenditures, it is still reasonable to assume that a part of this investment may have to come from the government. And therefore, given that there is constrained budget of the government, a higher expenditure in one sector, in this case the power sector, may result in a reduction in expenditures in other sectors. This is shown by Eq. (6.6):

$$I^B = \sum_i \Delta GE_i^A + \sum_i \Delta CF_i^A + \phi_g \Delta I + \left(1 - \phi_g\right)\Delta I \qquad (6.6)$$

where $\Delta I =$ total new exogenous injection into the economy, which is equal to the new capital required in the power sector, $\Delta I = \Delta CF'_{i\,=\,power}$. Φ_g is the fraction of total investment that comes from the government budget. The investment that is made by the government which presumably works within a specified budget then has to be sourced from other expenditure heads. Therefore this additional term of $\Phi_g \Delta I$, which is the additional investment that the government makes in the power sector, has to be deducted from other sectors of the economy if we assume that the government works within a fixed and constrained budget at any particular point in time. Equation (6.6) is therefore modified to give $I^{B\prime}$ which is in effect $I^B - \Phi_g \Delta I$. Equation (6.7) represents the operational form of this calculation:

$$I^{B'} = I^B - \phi_g \Delta I = \sum_i \left(GE_i^A - \phi_{ig} \Delta I \right) + \sum_i CF_i^A + \left(1 - \phi_g \right) \Delta I \qquad (6.7)$$

where Φ_{ig} is the fraction of government expenditure on the power sector that is deducted/sourced from sector i. Here $\sum_i \varnothing_{ig} = \varnothing_g$.

The next key variable therefore is the source of this additional government investment—i.e., how the government chooses to distribute its expenditure. For example, does welfare expenditure get reduced as a result of increased demands from the power sector or do all sectors get affected equally. The potential ranges for the values of the two key variables (Φ_g and Φ_{ig}) are 0–100%. The additional expenditure required in the power sector can theoretically come entirely from the government or entirely from the private sector or partially from both. Also, the way in which the government decides to source its share of the additional investment can vary between 0 and 100% for each sector, the total adding up to the total investment made by the government.

The investment required in the power sector is an output of Model 2 which evaluates the best (most cost effective) way to supply energy under multiple constraints of emissions, resource limits, capital, technical performance, etc.. This model is discussed in detail in Chap. 4. Multiple scenarios can be built using this optimization framework for different constraints on emissions, fuel supply options, fuel technology portfolio requirements, balancing requirements, etc. A range of potential values for some of the parameters and variables that play an important role in determining the fuel mix are provided in a later table. Some of the key parameters are the discount rate, fuel price escalation rate, and rates of reduction in price due to technology learning curves. The ranges for these three parameters for different countries and different technologies can be different. For example, the discount rate typically used for developed countries is different from the one used for less developed countries as for the latter the future is much less important in relation to the present. Also, the technology learning curves and the benefits accruing from the same for newer technologies such as solar and wind are much higher than those for older conventional technologies such as coal and hydro. The range of values for these parameters is obtained from a review of other literature—models, studies, and government reports—that uses these parameters to model future technology growth and market penetration.

The output from this optimization model is a mix of energy technologies that can supply energy over a certain period of time. The total investment required in the power sector and the effective cost of power generation are also the outputs of this Model 2. The value of total investment required in the power sector is then fed back into Model 3, the I-O model. This investment is the total new capital required in the power sector between the base and target year in Model 3 as discussed previously. For this new value of capital formation, a new economy-wide distribution of resources and production of goods and services is estimated in Model 3 among which is also a value for a new electricity demand. This electricity demand is compared with the value of the electricity demand in the previous iteration. If there is a difference of more than 10% in the two values, the entire process is repeated. The iterations continue till the difference between the values of electricity demand in consecutive iterations is less than 10%.

The linkage between Model 1 and Model 3 in the IMF is static, in the sense that there is no feedback loop between the two. There is however a check at each stage that the values assumed by input and output variables in Model 3 are within the bounds specified by Model 1. Model 2 and Model 3 however are dynamically linked in that the value of cumulative investment in the power sector between the base and target year, which is an output of Model 2, is an input in Model 3. Based on this and other exogenous inputs, Model 3 gives the values for production of goods and services in all sectors as an output. Model 3 also provides as outputs estimates of economy-wide energy demand and energy demand by each household class at the energy prices specified based on the output of Model 2. These are in turn fed back into Model 2 as inputs.

The range for the values of variables "energy intensity of GDP" and "sectoral GDP growth" is shown in Table 6.1. The changes in value added and energy intensity of GDP for 5-year periods starting from 1971 to 2005 were used to arrive at estimates shown in Table 6.1.

A range for potential values for each of the parameters and variables in the power sector model are given in Table 6.2.

Table 6.1 Range for sectoral growth rates and change in energy intensity of GDP in each sector from decomposition analysis of emission indicators between 1971 and 2008

		Minimum (%)	Maximum (%)
GDP growth (%)	Agriculture	1.60	4.60
	Industry	3.20	8.50
	Services	4.20	8.50
	Overall	3.50	7.10
Change in energy intensity (%)	Agriculture	0.50	9.00
	Industry	−5.20	0.60
	Services	−5.30	−1.30

Table 6.2 Range of potential values for parameters and variables in the power sector model

	Coal	Natural gas	Nuclear	Hydro	Renewable energy sources
Rate of reduction on price due to technology learning curves—minimum (%)	0.25%	1%	0.20%	0.25%	2%
Rate of reduction on price due to technology learning curves—maximum (%)	1%	3%	4%	1%	5%
Amount of new capacity that can be added in Year 1—minimum (GW)	15	5	3	3	8
Amount of new capacity that can be added in Year 1—maximum (GW)	0	0	0	0	0
Rate at which capacity addition changes (%)	5%	5%	5%	2%	5%
Discount rate—minimum (%)	4%				
Discount rate—maximum (%)	12%				
Fuel cost escalation rate—minimum (%)	1%				
Fuel cost escalation rate—maximum (%)	4%				

Data source: EIA (2013), McKinsey (2007)

The values shown in Table 6.2 are obtained from a review of literature including government reports[1,2] and reports of independent agencies that work in the energy sector[3,4] as well as from an analysis of past data for capacity addition.

6.2 Illustration of the IMF Through Scenarios

To demonstrate the operation of the IMF, a set of six key scenarios are constructed. Based on the range for growth rates in each sector obtained from Model 1, two main scenarios were constructed—scenario SS, in which economic growth is dominated and primarily driven by the service sector, and scenario SI, in which economic growth is primarily driven by growth in the industrial sector. The values for ΔG_i—GDP growth in sector i and ΔEI_i—change in the energy intensity of GDP in sector I for each scenario are given in Table 6.3. The overall growth rate in both scenarios is 7.1%, which is similar to the long-range average of the past two 5-year plans.

In addition to these, two scenarios are constructed in Model 2, the power sector model—Z_1 and Z_2. Keeping all other parameters the same, the only parameter that is varied between scenarios Z_1 and Z_2 is the constraint on emissions. In scenario Z_1, the

---

[1]Integrated Energy Policy, Planning Commission, Government of India 2006.

[2]Report of the Expert Group on Low Carbon Strategies and Inclusive Growth, Planning Commission, Government of India, 2011.

[3]International Energy Outlook, Energy Information Administration, USA, 2013.

[4]Pathway to a Low Carbon Economy, McKinsey, 2007.

Table 6.3 Values for ΔG_i and ΔEI_i for scenarios SS and SI—change between base year 2004 and target year 2030

		SS	SI
GDP growth ΔG_i (CAGR—%)	Agriculture	3.1	3.1
	Industry	5.8	8.5
	Services	8.5	7.2
Change in energy intensity of GDP (ΔEI_i—%)	Agriculture	0.5	0.5
	Industry	−2.3	−2.3
	Services	−3.3	−3.3

Table 6.4 Values for power sector parameters and variables for scenarios Z_1 and Z_2 for target year 2030

	Z_1	Z_2
Discount rates (%)	10%	10%
Fuel cost escalation rates (%)	2%	2%
Technology learning curve rates (%)	2%	2%
Base load requirement (% of total load)	30%	30%
Cumulative emissions (GtC)	No restriction	6

fuel mix is determined with no constraints on emissions or no requirements of a particular amount of renewable energy in the system. In scenario Z_2, an emission constraint of 6 GtC is imposed. India's INDC (Intended Nationally Determined Contributions) submitted to the UNFCCC commits to a 33–35% reduction in the emission intensity of GDP from 2005 levels by 2030. If we assume a 7% growth in GDP, this means that the emission intensity of GDP in 2030 has to be less than or equal to 0.03 tCO_2/Rs'000. If the annual emissions between 2005 and 2030 are calculated and cumulated, then this amounts to approximately 18 Gt of carbon at a GDP growth rate of 7% for India. Currently about 35–40% of total emissions can be attributed to the power sector. If this distribution remains constant between now and 2030, then it can be assumed that of the 18 GtC that India will emit, about 6 GtC will be from the power sector. As the INDC is a commitment that India has made towards global climate change mitigation, it can be considered to be a hard constraint on India's total emissions. A constraint of 6 GtC is therefore used in scenario Z_2 as a constraint on emissions for the power sector. The choice of values for the parameters and variables for both scenarios is shown in Table 6.4.

The scenarios constructed in Model 3 are connected to those in Model 2. For the particular fuel mix in each scenario Z_1 and Z_2, there will be an associated investment in the power sector. In scenario Z_2, the investment required in the power sector will be higher because of the emission constraint. Scenarios constructed in Model 3 capture the variations in the way in which this higher investment is made. In scenario X_1, it is assumed that of all the additional expenditure required in the power sector, 50% comes from the government and 50% is additional investment in the economy by the private sector. It is also assumed that the government sources this extra investment in the power sector from all other sectors in proportion to the

Table 6.5 Values of Φ_g and Φ_{ig} for scenarios X_1 and X_2 for target year 2030

		X_1 (%)	X_2 (%)
Additional expenditure in the power sector	Government share (Φ_g)	50	50
	Private share $(1 - \Phi_g)$	50	50
Contribution of each sector to govt. sourcing for additional expenditure	Agriculture $(\Phi_{ig})_{i\ =\ \text{agriculture}}$	21	0
	Industry $(\Phi_{ig})_{i\ =\ \text{industry}}$	28	0
	Services $(\Phi_{ig})_{i\ =\ \text{services}}$	51	100

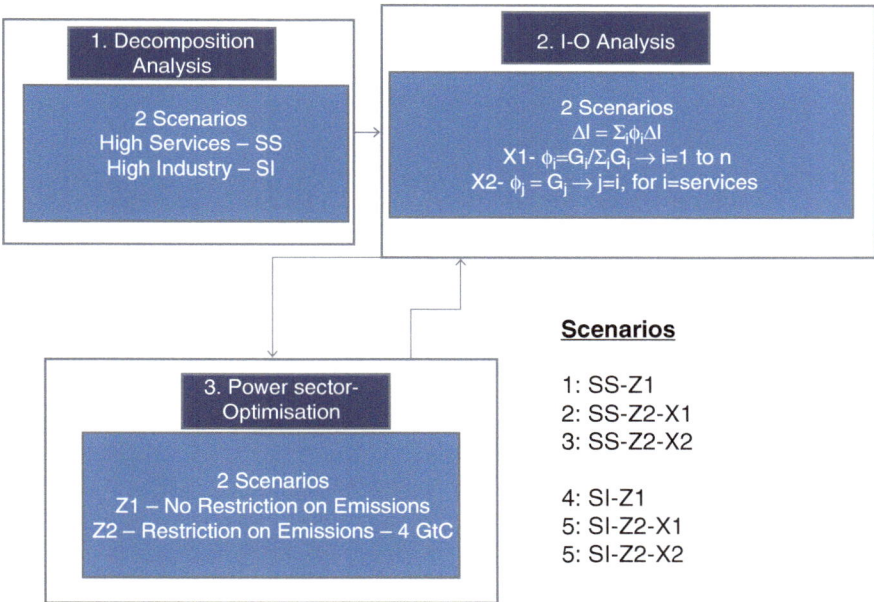

Fig. 6.2 Scenarios constructed using the integrated modeling framework for year 2030

sector's current (base year) contribution to GDP. In scenario X_2, the first assumption of 50% government expenditure remains the same; however the government now sources this additional investment by reducing welfare expenditure in key sectors. The scenarios X_1 and X_2 are shown in Table 6.5.

The six scenarios generated by a combination of two scenarios from each model are shown in Fig. 6.2

Scenario SS-Z_1 represents the scenario in which the growth in the economy is driven by growth in the service sector. There is no constraint on emissions in this scenario. Since the supply of energy is not constrained by a requirement of restricting emissions, no additional exogenous injection of investment is required in the power sector, i.e., in the economy; hence scenario X_2 of the I-O analysis is not relevant for this scenario. In the second scenario—SS-Z_2-X_1—while growth in the service sector continues to drive economic growth, now there is a restriction on emissions, which leads to a requirement of higher investments in the economy. The component of this additional investment that comes from the government is sourced from all sectors in proportion to the sector's contribution to GDP in this scenario. In the third scenario—SS-Z_2-X_2—while the first two components remain the same, the additional funding required is sourced from only the key welfare sectors, aggregated to the service sector in this analysis. The same sequence is then repeated for the set of scenarios constructed for GDP growth driven primarily by growth in the industrial sector—SI-Z_1, SI-Z_2-X_1, and SI-Z_2-X_2.

6.3   Results for IMF Scenarios

For ease of comparison, a few variables have been chosen across all scenarios and are presented in the following tables. Table 6.6 shows the values for economic and energy variables for the base year and scenarios 1 and 4, i.e., SS-Z_1 and SI-Z_1, the high services and high industry scenarios without any constraints on emissions.

In both these scenarios the overall growth rate is the same at 7.08%. There is also little difference in the values for total income of all household classes in both the scenarios. A larger variation is however observed in the values for incomes of particular household classes between the two scenarios. This is borne out by the GINI coefficient calculated for both scenarios. In both cases, the GINI coefficient has actually increased as compared to the base year.[5] This can be attributed to the fact that in this analysis, while the structural linkages in the SAM that determine the incomes in the target year remain the same, the population growth for each income class, that is given exogenously, changes. In scenario 1-SS-Z_1, the coefficient increases to 0.531 while in scenario 4-SI-Z_1, it increases to 0.527. This implies that while the overall GDP growth and overall household incomes remain more or less the same in both scenarios, scenario 4-SI-Z_1 implies more equitable growth as

---

[5]The GINI coefficients seen here for the base year, calculated using data from the 2003–2004 SAM constructed by Saluja and Yadav, are higher than the GINI's calculated and reported generally in government reports and by the World Bank. This is because due to the lack of any income data for India, more government surveys, on which World Bank also bases its calculation, use expenditure as a proxy for income. Now since expenditure levels saturate beyond a certain point, there is a high chance of undercalculating the actual inequalities among income classes and this is borne out by the low numbers seen in government and World Bank reports. However, in this calculation, the income of each income class is considered by adding to their expenditure also values for savings, and investments. Although these values are estimated from different RBI and NSSO reports while constructing the SAM, since they consider aspects beyond just consumption expenditure of household classes, the values for the GINI are higher. A comprehensive discussion of this issue can be found in Swaminathan and Rawal (2011).

Table 6.6 Income, electricity demand, investment, and equity in scenarios 1-SS-Z_1 and 4-SI-Z_1

	Base year	1—SS-Z_1	4—SI-Z_1
Per capita income (Rs.'000 /person/year)	22	88	88
GINI coefficient	0.497	0.531	0.527
Per capita electricity demand (kWh/person/year)	634	1919	2318
Total installed capacity (GW)	168	603	729
Annual investment in electricity sector (Rs. billion)		960	1234

Table 6.7 Energy variables for scenarios 1-SS-Z_1 and 2-SS-Z_2-X_1

	Base year	1-SS-Z_1	2-SS-Z_2-X_1
Per capita electricity use (kWh/person/year)	634	1919	1848
Total installed capacity (GW)	168	603	723
Installed capacity of renewable energy sources as percentage of total installed capacity (%)	13%	10%	32%
Annual investment in the electricity sector (Rs. billion)		960	1355
Per capita emissions (tCO$_2$/person/year)	1.66	4.61	3.74

compared to scenario 1-SS-Z_1. This is because, while the highest and lowest income classes earn a lower income in case of scenario 4 as compared to scenario 1, the income of the middle income groups, i.e., income classes RH3, RH4, and UH3, actually increases more under scenario 4-SI-Z_1 as compared to scenario 1-SS-Z_1. The per capita electricity demand, total installed capacity, and annual investment required in the power sector are all higher in the high industry 4-SI-Z_1 scenario than in the high services 1-SS-Z_1 scenario.

Now, let us take scenarios 1-SS-Z_1 and 2-SS-Z_2-X_1. These are both scenarios in which the initial sectoral growth rates are driven by high growth in the service sector. In these scenarios, the only difference is an additional constraint on total emissions that can be emitted from the power sector between 2004 and 2030. The requirement of restricting emissions results in a change in the fuel supply scenario. The component X_1 in scenario 2 indicates that the fraction of extra investment required for a higher deployment of renewable energy that comes from the government is sourced from all other sectors in the proportion of their contribution to GDP. The difference in the energy indicators between the two scenarios is shown in Table 6.7 and difference in the economic indicators for the two scenarios is shown in Table 6.8.

The per capita emissions reduce by 19% in scenario 2 as compared to scenario 1, however at the cost of a 41% increase in the total investments required. This increased investment is a result of the increase in the contribution of renewable energy sources in scenario 2 as compared to scenario 1. The impact of this increased investment on other aspects of the economy is summarized in Table 6.8.

Table 6.8 shows that as compared to scenario 1-SS-Z_1, there is a reduction in GDP growth in scenario 2-SS-Z_2-X_1. There is a reduction in the total incomes of households as well as an increase in the GINI coefficient in scenario 2 as compared to scenario 1. This is a result of an increase in the expenditure in the power sector.

Table 6.8 Comparison of economic variables for base year and scenarios 1-SS-Z_1 and $2\text{-SS-Z}_2\text{-X}_1$

	Base year	1-SS-Z_1	$2\text{-SS-Z}_2\text{-X}_1$
GDP (Rs. trillion)	27	157	153
GDP growth p.a.	–	7.08%	6.98%
Per capita income of all households (Rs.'000/year/person)	22	88	85
Per capita income of household class RH1	4	12	10
Per capita income of household class RH4	23	108	104
Per capita income of household class UH1	5	7	6
Per capita income of household class UH3	21	80	77
GINI coefficient	0.497	0.531	0.536

Table 6.9 Results for scenarios 1-SS-Z_1, $2\text{-SS-Z}_2\text{-X}_1$, and $3\text{-SS-Z}_2\text{-X}_2$

	1—SS-Z_1	$2\text{—SS-}Z_2\text{-X}_2$	$3\text{—SS-}Z_2\text{-X}_2$
GDP (Rs. trillion)	157.2	153.2	153.1
GDP growth p.a.	7.08%	6.98%	6.97%
Per capita income of all households (Rs.'000/year/person)	88	85	85
GINI coefficient	0.531	0.536	0.536
Per capita electricity use (kWh/person/year)	1919	1848	1883
Total installed capacity (GW)	603	723	737
Installed capacity of renewable energy sources as percentage of total installed capacity (%)	10%	32%	32%
Annual investment in the electricity sector (Rs. billion)	960	1355	1388
Per capita emissions (tCO$_2$/person/year)	4.61	3.74	3.91

Given a constrained budget of the government, this extra investment has to come from other sectors which in turn has a negative impact, albeit at these scales of renewable energy deployment a small one, on GDP growth, household incomes, and equity. Scenario $3\text{-SS-Z}_2\text{-X}_2$ is constructed to show that if the government decides to source this extra investment differently from different sectors, and not from all sectors in the proportion of their contribution to GDP as in scenario 2, the impacts may be different. Table 6.9 shows a comparison between all three scenarios, 1, 2, and 3 (1-SS-Z_1, $2\text{-SS-Z}_2\text{-X}_1$, $3\text{-SS-Z}_2\text{-X}_2$).

There is a minor difference between indicators calculated for scenarios 2 and 3 in this case, as this analysis is carried out on a highly aggregated SAM. A more disaggregated analysis to evaluate the impact of expenditure on, say, health and education in each scenario may provide better insights into the impacts of certain energy policies on certain income groups. However, the point of this exercise is not to arrive at policy prescription in terms of numbers for, say, emission or energy use, but to demonstrate that many scenarios can thus be constructed that can provide insights into potential trade-offs associated with certain governmental decisions and the ways in which these can be addressed.

Table 6.10 Comparison between scenarios 2-SS-Z_2-X_1 and 5-SI-Z_2-X_1

	2—SS-Z_2-X_1	5—SI-Z_2-X_1
GDP (Rs. trillion)	153	152
GDP growth p.a.	6.98%	6.96%
Per capita income of all households (Rs.'000/year/person)	85	84
GINI coefficient	0.536	0.532
Per capita electricity use (kWh/person/year)	1848	2233
Total installed capacity (GW)	723	874
Installed capacity of renewable energy sources as percentage of total installed capacity (%)	32%	32%
Annual investment in the electricity sector (Rs. billion)	1355	1712
Per capita emissions (tCO$_2$/person/year)	20.14	26.01

Table 6.11 Comparison of results for scenarios 1 and 4 with other modeling studies

	Model type	GDP growth rate (%)	Commercial energy use in 2030 (mtoe)	Per capita GHG emissions in 2030 (tCO$_2$/p)
NCAER	CGE model	8.84	1087	2.8
IRADE-AA	CGE—activity analysis model	7.66	1042	2.9
TERI-Poznan	MARKAL model	8.20	2149	5.0
TERI-MoEF	MARKAL model	8.84	1567	3.4
IMF-SS-Z_1—Sc-1	IMF	7.08	1688	3.6
IMF-SI-Z_1—Sc-4	IMF	7.08	2280	4.9

Similar results are obtained from the next set of "high industry" scenarios as well. A comparison of the impacts of a higher deployment of renewable energy in the high services scenario 2-SS-Z_2-X_1 and high industry scenario 5-SI-Z_2-X_1 is shown in Table 6.10.

Table 6.10 is presented here to illustrate that a policy of higher deployment of renewable energy will have a different impact if the baseline growth is dominated by service sector growth as in scenario 2 or industrial growth as in scenario 5. The results presented here illustrate the range of possibilities that can be explored using the IMF. The IMF therefore presents a way in which to evaluate multiple impacts, trade-offs, and costs of different decisions taken at the national level. Table 6.11 shows a comparison of the results of the IMF with those from other models. The model results compared in the table are the ones published by the Climate Modeling Forum published by the Ministry of Environment and Forests, Government of India, in 2010.

The assumptions in each model are different as is the structure of each model. The results for GDP growth vary widely across models as each model arrives at this value using a different method. While in the NCAER and IRADe models GDP growth is an output, calculated using different forms of the production function, in the TERI models it is an input that is used to then determine the amount of energy that would be required. In the IMF scenarios, the GDP is calculated based on the sectoral GDP values calculated as part of each scenario. The values for commercial energy use and emissions are also different in each model. This is because each model assumes a different rate of improvement in efficiency as well as a different sector-wise economic growth. Therefore, while the results are not directly comparable, they nevertheless provide a basis to judge the validity of the results obtained using the IMF.

6.4   Strengths and Weaknesses of the IMF

As stated earlier, the IMF represents only one possible combination of approaches that can provide a more comprehensive view of 3E linkages as an illustration of a different approach to 3E modeling. While there are advantages, there are however also limitations in using these specific modeling approaches. For example, the robustness of results from the input-output analysis is determined by the quality of data that is available to construct the input-output matrix in the first place. Beyond this, while an input-output analysis provides the structural linkages between sectors for a particular year, when this matrix is used to forecast values of production for the future, it is unclear whether these linkages will remain the same. It is also extremely difficult to predict the changes in these linkages for the future. The analysis therefore is subject to many assumptions and cannot be used for long-range forecasts. The problem of high uncertainties persists even in the optimization exercise as the actual results, even in very robustly specified models, can be subject to many uncertainties, both natural and epistemic. Also, there are many assumptions and exogenous "fixes" involved in making optimization models function smoothly. The model results can therefore be very sensitive to certain input parameters and therefore a robust sensitivity analysis for such optimization models is necessary.

However, despite these limitations, the framework proposed here is advantageous in providing insights into the trade-offs associated with various decisions regarding energy use and supply. Using different modeling methodologies in an integrated manner (as in the IMF) enables the user to address multiple questions relating to energy, climate, economy, and development together. The IMF allows the user to address aspects of poverty alleviation and equity among different income classes along with the technical and financial aspects of energy supply. In other modeling methods, either the CGE variety of models or the energy balancing and optimization models of the MARKAL variety, only one set of parameters can be addressed at any given point of time. The IMF also allows for the possibility of detailed modeling for the energy sector which is necessary to appreciate the actual costs and implications of high renewable energy deployment. Economic models often deal with aggregated production functions and detailed modeling of, say, the power sector as done here in the IMF is not possible in those models. A detailed energy sector model allows us to capture the nuances of the sector that can be vital in determining whether or not a planned level of deployment for a technology would be possible. The method of input-output analysis used here, viz., structural path analysis, allows the user to trace the way in which the effect of a particular investment or divestment will pass through the economy. This provides a way to analyze the impact of an increased investment in the power sector on other sectors of the economy and on household incomes and expenditures. So the linking of the energy sector model with the input-output analysis can provide a quantitative basis for analyzing the economic and developmental trade-offs associated with green energy policies. There are also disadvantages, especially in the use of the input-output method, over a regular general equilibrium model with production functions—the use of fixed multipliers and the lack of consideration to issues of scale being the most important ones. However, this method is favored here, especially in the case of analyzing scenarios for developing economies, as it not only requires a lower number of normative assumptions, but also allows greater flexibility in answering multiple sets of questions related to the linkages between emissions, energy use, and economy in a dynamically changing system.

6.5 Conclusions

The results obtained for the range of scenarios constructed here show that for a given set of assumptions, constraints, and structural linkages, the impacts of a high deployment of renewable energy on the economy and on incomes and income distributions are negative. The results also point towards the fact that the distributional impacts of a green energy policy may be different for different baselines—i.e., if the economy develops with an emphasis on manufacturing or industry or whether it develops with a service sector focus. The results point towards exercising caution in assuming that development and deployment of renewable energy technology will automatically lead to increased social well-being, as some studies want to do. There can be significant trade-offs involved in pursuing one goal versus the other and it is necessary to consider all the linkages that contribute to mitigating the negative impacts of increased government investment in renewable energy.

The IMF provides a way in which the values of energy and economic and environmental variables and their interactions with each other can be evaluated. The decomposition analysis provides a robust understanding of the variability in key control parameters, i.e., energy intensity of GDP and sector-wise contribution to economic growth. However, while it provides a sector-wise analysis of these variables, it does not provide a way to analyze the interaction and interdependence of these sectors. On the other hand, the input-output analysis has this advantage of capturing the inter-sectoral linkages. It also allows the user to evaluate the impacts of policy decisions and changes in control variables on important developmental parameters such as incomes and equity. However, as it is completely explicated in value terms, it cannot capture the physical realities of sectors such as the power sector. This aspect is then addressed by the third component which is a separate model for the power sector which takes into account the particular technical and economic characteristics of the sector and provides physical estimates for energy and emission variables. Therefore while each model is designed to address specific questions, used together in the IMF they provide a more comprehensive view of the interlinkages between the energy sector, economic system, and environmental constraints.

References

EIA, US. (2013). *International energy outlook 2013*. Washington: US Energy Information Administration (EIA).

GoI. (2006). Government of India, Office of the Principal Scientific Adviser and The Energy and Resource Institute. National Energy Map for India: Technology Vision 2030. New Delhi: TERI Press.

McKinsey, G. I. (2007). *Pathways to a low-carbon economy. Global greenhouse gas abatement cost curve*. Stockholm: McKinsey & Company.

Swaminathan, M., & Rawal, V. (2011). Is India really a country of low income-inequality? Observations from eight villages. *Review of Agrarian Studies, 1*(1), 1–22.

Taylor, L., & Taylor, L. (2009). *Reconstructing macroeconomics: Structuralist proposals and critiques of the mainstream*. Cambridge, MA: Harvard University Press.

Chapter 7
Conclusions

Abstract This chapter summarizes the main results of the study and re-emphasizes the need to use an integrated modeling framework instead of relying only on one modeling approach, especially in the context of developing countries when multiple, often conflicting goals are nevertheless equally important for policy. It draws from the previous chapters to conclude that while it is important to assess the best options of fuel technologies for a given energy system under a set of constraints, the impact of these choices on incomes and equity is an important factor that policy makers would need to consider while making decisions. Thus, if these questions have to be addressed simultaneously, it is necessary to integrate the methods needed to answer these. The integrated modeling framework therefore evaluates different aspects of the interlinkages between energy and economic systems. This chapters also discusses briefly the potential directions in which this research can be developed further to include more sectors as well as how it can be relevant to other countries in positions similar as that of India.

Keywords Integrated modeling framework · Equity · Climate change · Developing economies

The main contribution of this work is the development of an integrated modeling framework (IMF) that links three modeling methods. This framework addresses the research questions that were posed in Chap. 1.

The first question was this: Which factors are likely to drive energy requirements and emissions in a developing economy? How can their impact be estimated? This was addressed by a decomposition analysis of the emission intensity of GDP. This indicator was decomposed into the following key factors—(i) energy efficiency and level of modern energy use in a sector given by "energy intensity of GDP," (ii) the contribution of different fuel technologies in energy supply given by "emission intensity of energy," (iii) the proportionate contribution of each sector of the economy to the total GDP given by the ratio of sectoral GDP to total GDP, and (iv) the trends in total GDP. The results provide insights into the proportionate

© The Author(s), under exclusive licence to Springer Nature Switzerland AG 2020   105
T. Kanitkar, *An Integrated Framework for Energy-Economy-Emissions Modeling*,
SpringerBriefs in Environmental Science, https://doi.org/10.1007/978-3-030-18263-2_7

contribution that each indicator makes to the total change in the emission intensity of GDP over time.

The second question was as follows: How can these requirements be met under multiple constraints? What impact will, say, environmental constraints have on fuel supply options in a capital-scarce country? This was addressed by constructing a technologically explicit model for the power sector which estimates an optimal fuel mix, to meet a given energy demand, under multiple constraints. The mix of fuel technologies in an energy supply system depends on many factors. The availability of resources is one such factor. Some countries may, for example, have an abundance of hydropower potential while others may have vast oil or natural gas reserves. The characteristics of the load profile are another factor. For example, in regions where there is a higher energy demand from heavy manufacturing, the load profile will tend to look more flat as compared to a region where energy demand from the commercial sector is higher. In addition to these factors, there are economic factors that affect the choice of fuel technologies. Less developed countries, for example, have severe budgetary constraints that determine their expenditures in the energy sectors. The more evolved energy systems may also include considerations of energy security and diversifying the energy supply basket so as to insure themselves against uncertainties of global energy markets. In addition to these considerations, in the era of climate change, countries may also consider plans to include certain amounts of renewable energy sources in their energy baskets. These considerations along with the inherent technical characteristics of a particular energy system form the basis on which the energy system develops, adding new and decommissioning some old capacity each year. In order to capture these technical and financial details of the power sector, a model was constructed using GAMS (General Algebraic modeling system) which minimizes the discounted total system energy cost subject to multiple constraints, viz., resource potentials, emissions, base load requirements, system variability, and balancing potential. The construction of the power sector model using GAMS allows us to capture details of the sector while keeping the specification of the model relatively simple. It also allows us the flexibility to modify the objective function to include fixed and variable costs when needed, depending on the question that we are trying to address. For example, if we want to construct a merit order dispatch, then it is important to minimize the variable cost of energy. On the other hand if we are functioning within a fixed financial constraint, it may be important to minimize the fixed cost.

The third question was this: How can one evaluate the potential impacts of certain energy and economic choices on economic growth, incomes, and equity? This was addressed using input-output analysis. A social accounting matrix (SAM), which is an extended input-output matrix, was used as the basis for constructing an I-O model. The multipliers calculated for the base year represent the structural linkages between various sectors of the economy. These were used to determine the impact of a particular investment or divestment in the economy. The method of analysis used for this purpose is called "structural path analysis" and was developed by Defourny and Thorbecke in 1984.

The three questions that are addressed in this work are linked with each other. While it is important to assess the best options of fuel technologies for a given energy system under a set of constraints, the impact of these choices on incomes and equity is an important factor that policy makers would need to consider while making decisions. Thus, if these questions have to be addressed simultaneously, it is necessary to integrate the methods needed to answer these. The integrated modeling framework developed here therefore evaluates different aspects of the interlinkages between energy and economic systems. The impact of decisions taken in one domain on the other is estimated by interlinking the model that explicitly deals with the energy system with a model that represents the economic system in any region.

Operationally this means the following: the results from Model 1 which is the decomposition analysis are used to provide the upper and lower limits for the values of input and output variables of Model 3, which is the input-output model. Model 1 decomposes the trends in the emission intensity of GDP for time periods specified by the user and provides the operating limits for two key variables—(i) energy efficiency, i.e., energy intensity of GDP, and (ii) structural composition of the economy, i.e., values for sectoral GDP growth. In Model 3, the values for government consumption expenditure and capital formation, which are the exogenous sectors in the model, are adjusted in a certain ratio, till the GDP growth between the base and target year desired for a particular scenario is achieved as explained in the previous section. The value of total demand in each sector drives the demand for energy. This energy demand is to be met by a certain mix of fuel technologies determined through Model 2, which is the power sector model. Energy demand obtained from Model 3 is therefore an input for Model 2. Model 2 provides as outputs an optimal mix of energy supply technologies and the associated capital investment required in the power sector and unit cost of energy. These values for capital investment in the power sector and cost of energy are fed back into Model 3 and new values for economy-wide demand for goods and services are computed.

The results obtained for the range of scenarios show that for a given set of assumptions, constraints, and structural linkages, the impact of a high deployment of renewable energy on the economy and on incomes and income distributions are negative. This means that for a higher deployment of renewable energy sources, there is a reduction in GDP and household incomes. The results also point towards the fact that the distributional impacts of a green energy policy may be different for different baselines—i.e., the effort required to reduce emissions will be very different if the economic growth is driven by the manufacturing sectors or whether it is driven by the service sectors. The results indicate the need to exercise caution in considering social welfare and renewable energy development as two aspects which are inherently positively linked, i.e., increase in one leads to improvement in the other. There can be significant trade-offs involved in pursuing one goal versus the other and it is necessary to consider all the linkages that contribute to allaying the negative effects of increased government investment in renewable energy.

By dynamically linking the power sector model and the I-O model, two objectives are achieved. The first is that a model that can consider the specific technical characteristics of a particular energy system is built so that the constraints that may be important in a particular region can be included in the decision-making process. The second is that, through the construction of the economic model, the distributive impacts of a particular decision in the power sector, on economic growth and incomes, can be quantified. While there have been attempts to understand linkages between environmental sustainability and social welfare, these linkages are very often not completely quantified. The IMF offers a method to do so.

Very-large-scale macroeconomic models that were essentially developed to provide projections over short duration projections are now used to provide long-range forecasts, because of current policy requirements. However, even if these models are built using robust data and represent the existing relationships between key energy and economic variables fairly accurately, to use them to provide long-range forecasts is fraught with uncertainty. Economic development and changes in the patterns of energy use and emissions are a function of policy. Although some drivers of these changes are deterministic, for others, constructing probability distributions is very difficult. For example, while it may be possible to establish a relationship between the price of energy and energy use and it may even accurately describe the real world in the short term, any change in subsidy policies, or regional energy policies vis-à-vis global energy markets, would be likely to substantially impact this relationship. Therefore, it is more meaningful to study the impact of changes in one set of variables on another set given a range of different scenarios, instead of using models to forecast energy use and emissions. More insights and therefore more relevant information can be obtained from using models to describe the effects of policy changes (as captured by a set of parameters) on certain key variables. Therefore, an integration of different modeling methods and the intelligent construction of scenarios across these methods are advantageous, in order to address some of the important questions in the energy-economy-environment domain.

The IMF also opens up many directions in which this work can be carried forward. Apart from analyzing multiple scenarios and exploring various possible ways in which decisions can be taken and can affect key economic and energy parameters, the IMF can be gradually expanded to include an end-use analysis of other sectors as has been illustrated for the power sector. For example, it may be possible to include an analysis of the energy use in the transport sector to evaluate different options. In this analysis, production in a few sectors, electricity, iron and steel, and cement, is considered in physical terms in addition to the analysis in the input-output model that happens in value terms. In these sectors, the conversion from value terms to physical terms is relatively easier as time series data for prices and production is available in the public domain. For other sectors, availability of data acts as a major constraint in carrying out analysis in physical terms. However more sectors can be included in the analysis as data becomes more openly and transparently available.

Less developed countries find themselves in a unique position where they have to address both their developmental imperatives and their responsibility towards the global and local environment. It is therefore necessary to develop modeling methods and tools that can specifically address these challenges. The integrated modeling framework is proposed as the first step in this direction as it provides a method of analyzing evolving economic structures, concerns of equity and income distributions, and impacts of climate constraints in an interconnected manner.

References

Defourny, J., & Thorbecke, E. (1984). Structural path analysis and multiplier decomposition within a social accounting matrix framework. *The Economic Journal, 94*(373), 111–136.

Further Readings

Allen, M. R., Frame, D. J., Huntingford, C., Jones, C. D., Lowe, J. A., Meinshausen, M., & Meinshausen, N. (2009). Warming caused by cumulative carbon emissions towards the trillionth tonne. *Nature, 458*(7242), 1163–1166.

Apergis, N. (2015). Does renewables production affect income inequality? Evidence from an international panel of countries. *Applied Economics Letters, 22*(11), 865–868.

Baer, P. (2013). The greenhouse development rights framework for global burden sharing: Reflection on principles and prospects. *Wiley Interdisciplinary Reviews: Climate Change, 4*(1), 61–71.

Bows, A., Mander, S., Starkey, R., Bleda, M., & Anderson, K. (2006). *Living within a carbon budget*. Manchester: Tyndall Centre.

Cancun Agreements. (2011). In United Nations Framework Convention on Climate Change. Retrieved January 22, 2015, from http://unfccc.int/resource/docs/2011/sb/eng/inf01r01.pdf.

Climate Change Synthesis Report, Fourth Assessment Report. (2007). Retrieved August 14, 2015, from http://www.ipcc.ch/pdf/assessment-report/ar4/syr/ar4_syr.pdf.

Climate Modelling Forum, India, Ministry of Environment and Forests, Government of India. (2009). *Climate Modeling Forum, India's GHG Emissions Profile—Results of Five Climate Modeling Studies*. New Delhi: India. Government Press.

Del Río, P., & Burguillo, M. (2009). An empirical analysis of the impact of renewable energy deployment on local sustainability. *Renewable and Sustainable Energy Reviews, 13*(6), 1314–1325.

Ekholm, T., Soimakallio, S., Moltmann, S., Höhne, N., Syri, S., & Savolainen, I. (2010). Effort sharing in ambitious, global climate change mitigation scenarios. *Energy Policy, 38*(4), 1797–1810.

Enkvist, P., Nauclér, T., & Rosander, J. (2007). A cost curve for greenhouse gas reduction. *McKinsey Quarterly, 1*, 34.

Experts, B. A. S. I. C. (2011). Equitable access to sustainable development: Contribution to the body of scientific knowledge. In *A Joint Paper By Experts from BASIC Countries*. Beijing, Brasilia, Cape Town and Mumbai: BASIC Expert Group Retrieved from http://www.erc.uct.?ac.za/Basic_Experts_Paper.pdf.

Fan, Y., Liu, L. C., Wu, G., & Wei, Y. M. (2006). Analyzing impact factors of CO2 emissions using the STIRPAT model. *Environmental Impact Assessment Review, 26*(4), 377–395.

IPCC. n.d. Fourth Assessment Report, Working Group III—Mitigation of Climate Change—Discount Rates. In Inter-Governmental Panel on Climate Change. Retrieved April 26, 2015, from http://www.ipcc.ch/publications_and_data/ar4/wg3/en/ch2s2-4-2-1.html.

Hsu, G. J., Leung, P., & Ching, C. T. (1988). Energy planning in Taiwan: An alternative approach using a multi-objective programming and input-output model. *The Energy Journal, 9*(1), 53–72.

Hubbert, M. K. (1975). Survey of world energy resources. *Energy Sources Future, 1*, 3–38.

Jayaraman, T., Kanitkar, T., & D'souza, M. (2012). Equity and burden sharing in emission scenarios: A Carbon Budget Approach, in Dubash, Navroz (ed.), Handbook of climate change and India: Development, politics and governance, Earthscan, New York, pp. 131–146.

Kanitkar, T., Jayaraman, T., D'Souza, M., & Purkayastha, P. (2013). Carbon budgets for climate change mitigation—a GAMS-based emissions model. *Current Science (Bangalore), 104*(9), 1200–1206.

Mackay, R. M., & Probert, S. D. (1993). Predicted supplies of oil: A controversy revisited. *Applied Energy, 46*(2), 113–147.

Marchetti, C., & Nakicenovic, N. (1979). The dynamics of energy systems and the logistic substitution model. Technical Report. PRE-24360.

Martinez, D. M., & Ebenhack, B. W. (2008). Understanding the role of energy consumption in human development through the use of saturation phenomena. *Energy Policy, 36*(4), 1430–1435.

Matthews, H. D., Gillett, N. P., Stott, P. A., & Zickfeld, K. (2009). The proportionality of global warming to cumulative carbon emissions. *Nature, 459*(7248), 829–832.

Meinshausen, M., Meinshausen, N., Hare, W., Raper, S. C., Frieler, K., Knutti, R., et al. (2009). Greenhouse-gas emission targets for limiting global warming to 2°C. *Nature, 458*(7242), 1158–1162.

Meyer, A. (1999). The Kyoto Protocol and the emergence of "contraction and convergence" as a framework for an international political solution to greenhouse gas emissions abatement. In Man-Made Climate Change: economic aspects and policy options. Physica, Heidelberg, 291–345.

Miles, H. (1998). *The socio economic impact of renewable energy technologies (No. 92).* Surrey Energy Economics Centre (SEEC), School of Economics, University of Surrey.

Nilsson, S., & Schopfhauser, W. (1995). The carbon-sequestration potential of a global afforestation program. *Climatic Change, 30*(3), 267–293.

Oberheitmann, A. (2010). A new post-Kyoto climate regime based on per-capita cumulative CO_2-emission rights—rationale, architecture and quantitative assessment of the implication for the CO2-emissions from China, India and the Annex-I countries by 2050. *Mitigation and Adaptation Strategies for Global Change, 15*(2), 137–168.

Pradhan, B. K., Saluja, M. R., & Parida, Y. (2014). A social accounting matrix for India, 2005–06. *Research in Applied Economics, 6*(1), 176–201.

Sanders, I., Batty, W. J., Probert, S. D., Hagino, K., & Aida, S. (1993). Supply of, and demand for, a resource: Fuzzy logistical optimisation technique. *Applied Energy, 46*(4), 285–302.

Shukla, P. R. (1995). Greenhouse gas models and abatement costs for developing nations: A critical assessment. *Energy Policy, 23*(8), 677–687.

Smil, V. (2000). Perils of long-range energy forecasting: Reflections on looking far ahead. *Technological Forecasting and Social Change, 65*(3), 251–264.

Smil, V. (2008). Long-range energy forecasts are no more than fairy tales. *Nature, 453*(7192), 154–154.

The National Academy of Sciences; America's Climate Choices: Limiting the Magnitude of Future Climate Change. (2010). Retrieved April 26, 2015, from http://www.nap.edu/openbook.php?record_id=12785&page=21.

United Nations Development Report. (2015). Human development report. Retrieved May 5, 2016, from http://hdr.undp.org/en/data.

Weyant, J., Davidson, O., Dowlabathi, H., Edmonds, J., Grubb, M., Parson, E. A., et al. (1996). *Integrated assessment of climate change: An overview and comparison of approaches and results (pp. 367–396).* Cambridge and New York: Cambridge University Press.

Index

A

Arithmetic mean Divisia index (AMDI), 31–33
Autonomous energy efficiency improvement
　　(AEEI), 20
Autonomous energy efficiency index
　　(AEEI), 18

B

Baseline emissions, 30, 107
Bottom-up energy models, 13
Brundtland Commission, 3
Budgetary constraints, 106
Budget-constrained economy, 84, 85
Business-as-usual (BAU), 39, 80, 81, 83, 85

C

Carbon budget (CB), 53, 56, 60
Carbon capture and sequestration (CCS), 50
Central Electricity Regulatory Commission
　　(CERC), 50
Climate change, 1–3, 36, 63, 64, 96, 106
Computable general equilibrium
　　(CGE), 11, 21, 65

D

Decomposition analysis, 5, 6
　　actual and projected trajectories of
　　　　growth, 29, 30
　　agriculture, industry and services sectors, 28
　　AMDI, 31
　　baseline cumulative emissions, 40
　　business-as-usual baselines, 42

capital formation, 91
demand sectors, 91
developing economies, 28, 29
economy-wide energy demand, 91
emission and energy indicators, 34
emission intensity of GDP, 90, 105
energy and emission intensity (*see* Energy
　　and emission intensities)
energy demand, 91
energy efficiency, 29, 40, 42, 107
energy intensity, 34
exogenous adjustment, 92
factors, 30
fuel mix, 92
GDP, 40
global climate agreement, 40
government budget, 93
government expenditure, 91
government investment, 93
greenhouse gas emissions, 27
India, 31, 32, 34
LMDI, 31
mitigation effort, 40, 41
parameters, 30
planned government and private
　　expenditures, 92
sectors, 28
social accounting matrix, 91
structural composition, 30, 107
structural linkages, 92
target year 2030
　　developing countries, mitigation
　　　　actions, 39, 40
　　economic baselines, 38, 39
transitions, 29